A PRACTICAL GUIDE

ONLINE GALLERY

To view more beekeeping photos, as well as larger versions of the photos already included in this book, scan the QR code shown on the front cover with your mobile or tablet or simply enter **fug.io/bees** into your web browser.

ABOUT THE AUTHOR

Roger Patterson started keeping bees in 1963 as a 15-year-old after the hard winter of 1962/3. He became a committee member of his local Beekeepers' Association (BKA) at the age of 17, chairman at 23 and county BKA chairman at 25. At one stage he kept 130 colonies.

He is a practical beekeeper, concentrating on the basics and keeping things simple. He manages the local BKA apiary where there are usually a large number of colonies for instruction. He has been a demonstrator for well over 35 years, lectures widely and believes beekeeping should be fun. He is a British Beekeepers Association (BBKA) Trustee, Bee Diseases Insurance (BDI) Vice President and BIBBA (Bee Improvement and Bee Breeders Association) committee member.

In 2011 Roger took over the administration of Dave Cushman's website **www.dave-cushman.net**. This is considered by many to be one of the most comprehensive and authoritative beekeeping websites.

BEEKEEPING
A PRACTICAL GUIDE

Roger Patterson

RIGHT WAY

Constable & Robinson Ltd
55–56 Russell Square
London
WC1B 4HP
www.constablerobinson.com

First published by Right Way, an imprint of
Constable & Robinson, 2012

A copy of the British Library Cataloguing in Publication Data
is available from the British Library

ISBN: 978-0-7160-2285-5

Printed and bound in the EU

1   3   5   7   9   10   8   6   4   2

CONTENTS

ILLUSTRATIONS

by David Woodroffe

COLOUR PLATES

Photographs taken by the author, except for Plate 5, courtesy of Neville Childs, and Plate 7, courtesy of John Glover.

DEDICATION

I dedicate this book to the late George Wakeford BEM 1900–1985. He was born and brought up on a farm in West Sussex and stayed very close to the land all his life. Although he had minimal education he was incredibly knowledgeable about nature and farming. As with all practical people he had great stock sense that he applied to all animals including bees. Although he had never read a bee book in his life and didn't know much about the technical side of beekeeping he was the best handler of bees I have ever seen and by some considerable distance. He never wore a veil or gloves, yet could handle the most vicious colony in a calm and methodical way without getting badly stung.

He was a very quiet man and not a good teacher, but I learnt a huge amount simply by watching him. Many of my methods were formed in this way and I'm pleased I have been able to keep refining them and pass them on to others.

He was so well respected in West Sussex beekeeping he was awarded the British Empire Medal in 1981.

I was incredibly lucky in knowing George Wakeford well and I will always be grateful for the opportunity to learn so much from him.

ACKNOWLEDGEMENTS

I would like to thank Sue Cooper and Brian P. Dennis for reading the original draft. They are both beekeepers and although they made some useful suggestions about the beekeeping content they concentrated mainly on my "Sussex English" and punctuation (whatever that is) in an effort to help the reader understand what they thought I was trying to say. They have done a good job because even I can understand it now!

I would also like to thank Judith Mitchell, my editor, for her patience and almost instant response to emails. I found her comments invaluable in seeing the content from a non-beekeeper's point of view.

INTRODUCTION

This guide is written for those of you considering taking up beekeeping, or about to start out in the hobby or with just a little experience with bees. Its aim is to encourage you and help you to understand how to care for your bees successfully. Ultimately, I hope your experience will lead you to agree with my long-term view that beekeeping is fun!

I have been keeping bees since 1963 and I had 130 colonies for several years. Apart from my first couple of years of beekeeping I have always tried to improve my bees by rearing my own queens from local colonies and I encourage others to do so, rather than use imported ones. I have been teaching since the early 1970s, most of which has been at the teaching apiary of my local beekeeping association (BKA), the Wisborough Green Division of the West Sussex Beekeepers' Association.

I also lecture and demonstrate widely on practical beekeeping, so meet a lot of beekeepers both new and old and often come across some ideas that are real gems that people have used for years. I always look to see if a new idea will suit the way I work; if it doesn't, I either discard it or modify it. We never stop learning.

Most of the information I give is based on experience: firstly, I have actually done it in my own colonies; secondly, teaching has helped me to understand what beginners need in their formative years. In beekeeping you will never be short of advice either written or verbal, but you need to be aware that some of it is not based on experience, but what has been gleaned or cobbled together from other sources, often by inexperienced beekeepers.

Many people, even beekeepers, try to humanise bees and in my opinion that is wrong. A colony is a cohesive unit and unable to think in the same way that we do. We know they are largely controlled by pheromones, but we probably only know a fraction of what goes on inside a colony. There is a lot still to find out.

I was brought up on a farm, am an engineer by trade and from an early age I have been practical and capable of lateral thinking, assets I have found useful in my beekeeping. I am largely self-taught and by nature tend to keep things simple; that way I can usually understand why I have a problem and find a way out of it. In my teaching I try to impress upon beginners there are two very important things you must learn in order to become a good beekeeper: good handling techniques and learning what I call the "basics". By basics I mean factual things that very often many with several years' experience don't know. These include how to:

- Identify brood in all stages, pollen and honey stores.
- Learn the life cycles of the queen, drone and worker.
- Recognise good healthy bees and brood.
- Identify the brood diseases – American Foul Brood (AFB), European Foul Brood (EFB) and Chalk Brood – and know how to deal with them.
- Learn the life cycle and treatments for *Varroa destructor*.
- Understand how varroa vectors viruses.
- Understand the swarming process.

Don't worry about these at the moment, as you should pick up enough information if you read on.

In a couple of places in this book I have expressed some strong views that may be seen as controversial. I am passionate about bees and beekeeping and have seen what damage has been done to them, sometimes in the name of kindness. Those who know me well will tell you I have endless time for those who work hard to become good beekeepers and I get frustrated and angry when I see people given bad information and advice, some of which has crept into modern beekeeping teaching. I am happy for methods other than mine to be shown provided that they suit the system that is being advocated. However, often "factual" information is incorrect. Bees don't always respond in the way you expect, and two colonies on the same site at the same time that are treated the same will often react differently and things are not always as straightforward as they seem. You only learn these things by experience and you can't always get answers from a book or a phone call. Some may well be offended by this paragraph and I make absolutely no apologies. I consider my standards are high and I want all beekeepers to have access to good sound teaching, so they can look after their bees well and enjoy their beekeeping.

The modern beekeeper needs to be much more knowledgeable than beekeepers were in the past. There are introduced pests and diseases that need constant attention and knowledge to deal with them and the beginner must get up to speed very quickly.

I have always been confused as to what a beekeeping "beginner" is. A dictionary definition of beginner might be something like "a learner", but we are all learners. For the purposes of defining ability groups at Wisborough Green BKA, I expect those in the Intermediate group (which is the next one up from Beginner) to be able to handle a colony entirely on their own, know the basics and be able to deal with most things that are thrown at them throughout the season including:

• Dealing with a swarming colony
• Taking and hiving a swarm
• Making colony increase
• Clipping and marking a queen by hand or
• Be willing, eager and capable of learning any of these reasonably fast.

If they are unable to do any of these or similar things then they stay in the beginners' group, so perhaps we can use that as a yardstick. Some may think this is quite tough and it is, but beekeeping has got much tougher and there is now an urgency about learning. As a guide I would expect that someone who is practical and positive and who has had a good grounding from a good beekeeping association should consider him or herself a beginner for no more than a couple of seasons. I think it is a mistake to judge people on their length of time in beekeeping as many will quickly progress from the beginners' stage and have far more knowledge and ability than someone else who may have had bees for 10–20 years, something you may need to remember when looking for an "experienced" beekeeper, as there are a significant number of "perpetual beginners".

I have deliberately kept things simple and I know that the more experienced will question some of my advice. For example, I have suggested that when a colony is starving in winter or spring you should source candy from an equipment supplier. I know you can get it cheaper elsewhere or make it, but I have gone for the easiest option, rather than waste precious time that could result in a dead colony. Some of the advice is perhaps not what I would give a more experienced beekeeper, but it will get you out of trouble. With experience you will modify some of what I advise and I encourage you to do so, but hopefully you will look back and understand why I have recommended certain action.

I am in no doubt that an experienced beekeeper will point out some omissions but I have tried to include most of what a potential or new beekeeper (I hate "newbie" or "newbee") should need and have tended to concentrate on them rather than perhaps slightly more advanced topics that can be sourced elsewhere.

I have always thought that attempting a book to appeal to both a non-beekeeper and someone who has kept bees for one or perhaps two full seasons is incredibly difficult and I have resisted many suggestions to write one until now. I thought it would be a problem giving just enough information to non-beekeepers that would allow them to decide whether to go further without confusing them and enough for beginners to help them in their first season or two. A simple example is that someone who has only kept bees a few weeks might need to know in detail what to do when their colony has swarmed, yet a non-beekeeper would only need to know why the bees have swarmed. By nature that sort of book would only have a short period of usefulness, as it will never become a reference book. For that reason when you have finished reading this book I ask you to keep it and lend it to others who might like to take up this absorbing hobby.

The Perception of Beekeeping

I have given many talks to non-beekeeping organisations and spent many hours on stands at shows. There are very definitely two main groups of non-beekeepers. Some are already incredibly knowledgeable and ask sensible questions and these are often the young which is encouraging. I find it very rewarding to speak to these people, but others aren't quite in the same league. There are three words they seem to associate with bees: stings, swarms and honey. Some questions are predictable and I find it a bit difficult continuing a conversation if the first question is, "Do you ever get stung?" A long time ago I stopped counting how many times I'd been asked if I beehive myself or if I knew about the birds and bees!

The old-fashioned view of a beekeeper is a little old man or woman who hides away for hours performing some sort of magical trick with bees in order to produce a little honey. But modern beekeepers are no different from any other group of people. When I started beekeeping the vast majority of beekeepers worked on the land in some way, now very few do and the average age has got very much younger in a short time. Not long ago we had difficulty getting people to winter evening meetings because they wouldn't drive in the dark, now it's because their children have activities such as scouts, guides and school events. At the moment the occupations of some of our members at random are gardener, entomologist, architect, horticulturist, policewoman, engineer, farrier, accountant, plumber, etc, etc and I reckon you can find these people in any group.

An Appeal on Behalf of Bees

If you are thinking of making a present of a colony of bees to someone without their knowledge, please DON'T. You may think it will make a good wedding or retirement present, but if that person is more responsible than you are they will soon realise the huge commitment you have sentenced them to. By all means give a present of a book or perhaps enrolment in a beginners' course so they can make up their own mind, or ask them if they would like to keep bees, so they have a chance of thinking about it. Virtually every time I have known bees been given as a present without consent, it has resulted in disaster. If you gave an unwanted puppy and it was neglected it could result in prosecution, but unfortunately that doesn't happen with bees.

Why Do People Keep Bees?

I find it incredible that many beekeepers, even some very good ones who have kept bees a long time, don't actually like honey. So why do they start beekeeping? There are many reasons, such as:

To Pollinate the Garden

A hive of bees is unlikely to make much difference to a small garden, but it will to the surrounding area, as bees will normally fly around 1½ miles in search of food. Above that and it may not be economical for them to do so.

Honey for the Family

I suppose it depends on how big your family is. I'm sure the average family will soon find uses for it including in the kitchen, but any surplus could be used as presents, thank yous and bartering.

It's a Hobby

The vast majority of beekeepers treat beekeeping as a hobby and it is actually quite a good one. Most of it is performed in the open, there are always different things to see and learn, it is environmentally friendly and relatively inexpensive compared to other activities.

For the Income

There are some people who come into beekeeping thinking they are going to make a fortune. I think they remember hearing that someone's grandfather took 329 lb of honey from one colony in 1929, read of it being on sale at an exclusive outlet in London for £XX per lb or were told of a bee farmer who works 500 colonies on his own. Some of this is anecdotal and probably exaggerated but they still get the calculator out and start dreaming. They don't stay long. Even commercial beekeepers usually need something else in order to make a living. When I started beekeeping many beekeepers worked on the land and kept 20–30 colonies to augment their income, but that is a totally different matter as their overheads were low. There are now some people who have retired early and keep perhaps up to 100 colonies, but this is often simply an extension of their hobby and they may have other income.

Interest in Nature

Not only are bees incredibly interesting but there are many things to see both in and around the hive. I myself am interested in natural history and always encourage beekeepers to look inside a hive roof when they take it off. I have kept bees in some wild places and if I had time when I finished hive inspections I would sit down quietly for half an hour or so, watching with surprise as the surrounding area came to life.

Uses for Hive Products

Many think of honey as something you take when you have a cold, or to spread on bread, but it can be used in cookery and made into mead. Bees also produce beeswax and that can be made into candles, cosmetics, soap, etc. Of course these can all be made by non-beekeeping members of the family.

Further Interest

Beekeeping also gives you opportunities to develop other interests such as photography and microscopy.

1

HOW DO YOU LEARN?

Beekeeping is a vast subject, much of which the ordinary beekeeper doesn't need to know, but to keep bees successfully you will need to grasp the basics. You will come across much information and many methods of doing things, some of which will contradict others, some will complement them, some will be complete rubbish, some simple, some complicated, some will be very sound and you will have to sort the wheat from the chaff. If you simply try to put into practice what you are told without understanding it you may well come horribly unstuck, even lose bees or a lot of money because you have bought something that is useless, that you can't sell because everyone else knows what you don't. The sensible thing to do is to fully understand the principle and work out for yourself if it is feasible. Many make the mistake of simply following instructions, then when it goes wrong they can't work out why, or how to put it right.

I would expect a beginner to struggle to understand something that is fresh, so why not jot it down graphically in some way? How about doing it in "PowerPoint"?

With the current influx of new beekeepers there are a small number of more experienced beekeepers to teach them. This together with the fact that most beekeepers only keep a handful of colonies or fewer means that many teachers are quite inexperienced. This has unfortunately resulted in a mentality that thinks all you have to do is provide training material and anyone can teach or learn beekeeping. I have been involved long enough to know that beekeepers learn at different rates and have different levels of interest. It is a hobby and there is no compulsion to attain a certain level as there might be in a work situation. For those reasons I don't believe that a "one size fits all" approach is satisfactory.

Information and Misinformation

With my background I have always tended to question things I am told especially if they don't quite add up. I ran my own business for 25 years and on occasions that philosophy served me well, even though I was dealing with customers who were always supposed to be right and employees who were supposed to be on my side.

In recent years culture has changed and in whatever we do we are "trained" to do things rather than rely on the old-fashioned values of knowledge, logic, lateral thinking and common sense. This has found its way into beekeeping and there is a lot of dreadful twaddle both spoken and written about bees, some of which has mysteriously become "best practice", often being advised as such by quite inexperienced people.

In my view beginners should expect the information they are given to be correct and reliable. At the time of writing there is a huge upsurge of interest in beekeeping and I am seeing a lot of people who are new to the craft. Many are enthusiastic and hungry for knowledge. They may have already done some reading and in principle that sounds like a good idea, but when they are told that some of what they have read or heard is erroneous or bad advice they can be forgiven for being confused. What do they believe? Something that is written in several places or "ordinary" amateurs who are speaking from experience?

R.O.B. Manley was a very successful commercial beekeeper either side of World War Two and I believe was the first man in the UK to keep 1,000 hives. He wrote several books and in one he complained about exactly the same thing when he started beekeeping in the early years of the twentieth century, so we haven't solved much. There is some very good information available, but a significant amount of poor. Unfortunately you need to become experienced before you can decide which is which and in the meantime you can make many mistakes that aren't your fault. It is difficult to advise on how to deal with this unless you are in the lucky position of having good local tuition.

Some of what I write will be frowned upon by many, simply because it may not be found elsewhere, but I can guarantee it is all based on my own experience of nearly 50 years dealing with many more colonies than the vast majority of beekeepers, unless for the sake of completeness I have given information that has been observed or researched by others and experience tells me it is sound. Are you confused? Of course you are, but I'm on your side as I'm teaching people like you on a regular basis and know how difficult it is for you.

I will give you an example of the written word. Recently I looked at a beekeeping association's website and on a page entitled "Information for New Beekeepers" I came across all the following in different places on the same page, which I have simply cut and pasted: "the queen will lay 3,000 eggs a day", "there are 50,000 worker bees in a hive", "a worker bee lives for 8 weeks". Now one thing I'm reasonably good at is mental arithmetic and could easily see that 3,000 x 56 isn't 50,000. Even allowing for a reasonable number of casualties, at least one of these figures must be seriously wrong, but how many others have noticed it? Few I suspect because I have seen these or similar figures in many places, but how many who quote them have actually checked them? None I suspect. They have simply taken them as being correct because it's in print. There was no mention of the time of year, type of bee or the country, all of which will make a considerable difference. I have also come across two beekeeping associations who have got images of hoverflies on their website home pages instead of honey bees!

I know this example may seem trivial and it won't have any effect on beekeeping, but it is typical of some of the stuff that is taught by BKAs who you would expect to know better and there is plenty more where that came from. If that can't be believed, what can?

It is definitely not my intention to put anyone off bee-keeping, or learning about it. My sympathy goes to you folk who have to work out what is actually correct. Bees don't always do what they should and you will probably find that out at an early stage, but by understanding the basics you will be able to sort many problems out yourself.

I think beekeeping is a wonderful craft and it gives me a huge amount of enjoyment to see beekeepers at Wisborough Green or anywhere else doing well. We have a core of good members who have been beekeeping four years or less and in my opinion they have done incredibly well. They have had good quality tuition that has allowed them to study further and reject the duff information, but some people haven't had that opportunity. Please don't think that all information is bad, it certainly isn't, but I urge you to question what you are told if it sounds doubtful.

Practical Experience

There really is no substitute for this and it's probably the best way to learn. If you are a practical person who picks things up quickly you should be able to learn by watching others, even if it's how not to do something. When you have been to several meetings you will probably become friendly with like-minded people of a similar ability, to the point where you are able to work with them. You have a common interest and quite probably you will find you will learn more by talking and working together. Of course you will make mistakes; it's recognising and learning from them that's important.

A really good BKA will have a good and well run teaching apiary where demonstrators give tuition to others and will allow members to handle a colony entirely on their own. (Plate 1 shows a scheduled meeting at the Wisborough Green teaching apiary.) This in my opinion is the best way to learn as you have guidance and help if something goes wrong. Always take the opportunity to handle bees, as this is where your techniques will be developed. Watch every move of the others and if you are sharp you will learn a lot. Look at the movements of both handler and bees. If the bees get agitated or flighty, is it something the handler did or didn't do? Is the handler looking for disease, amount of space or food?

In my opinion it is just as important to see bad handlers as it is good ones. You will probably find that someone who is a rough handler with crash-bang methods will fire the bees up, yet the person who is calm and methodical may not have the same problem.

It is difficult to give advice on handling and inspecting without being at a hive, but every colony is different and the bees are telling you something all the time. The "reading" of a colony is extremely important and has to be learnt as it's very difficult to teach it. Knowing when to smoke the bees and how much is of vital importance.

Theoretical Knowledge
To be a good and successful beekeeper not only will you need to be good at a practical level, but you will need to know the theory to a reasonable depth as well. They must go together in modern beekeeping. This doesn't need to be in any great detail at the beginner stage, but continuous learning is a part of beekeeping. I am writing this the day after attending a lecture on "Forage and Navigation" where I learnt several things I didn't know before. That's after nearly half a century of beekeeping and although it won't help me look after my bees any better it will improve my understanding of them.

Books
Despite the presence of the web, books are probably still the first place a person will look for information and one area I have a serious problem with. I am regularly asked what book a beginner should read and quite frankly I have difficulty answering. There are very few really good books but I always make a couple of suggestions where the factual content is very sound. In general these books stand the test of time and are usually at the top of the list of most good beekeepers.

Some books are simply awful. Sadly there are some people with little or no knowledge who seem to think they know enough to write a book. Some of our members are amazed at an apiary meeting when I say, "I've never seen that before", as I often do. If I'm still learning, how can someone who has only been keeping bees a year or so, less in some cases, possibly be knowledgeable enough to write a book with any authority? Some books are factually incorrect and can't even get the terminology right, to the point where I often know which book someone has read by their comments and questions. The cynic in me suggests that due to their lack of experience the authors probably read other books and simply rehash what others have written and copy mistakes that have been copied before, simply because the writers don't know that the information is dubious. You then get the same erroneous information in several places; it gets read by a beginner and is taken as being correct. What else are beginners to do until they have enough experience to find out for themselves? In my opinion a person who buys a book on a specialised subject like beekeeping has the right to expect it to be written by an experienced person and the information to be sound, especially when we are dealing with a box full of potentially dangerous animals.

The lack of good books gives the beginner a problem. You need to know enough to know if the information is correct and if you know that amount you don't need the book in the first place. I still think books are important, so I suggest you learn the basics and with a bit of common sense and logic most other things should drop into place. Having done that you should be able to work out what is reasonably reliable and in any case you can check with a good beekeeper at your local BKA about which book to read and stick with it.

At the beginner stage I would completely ignore foreign books, especially American. Almost everything is different – climate, conditions, hive types, bee types, legislation, etc – and you could be seriously misled as many often are.

Some of the older books still have incredibly sound beekeeping content and are well worth reading, especially for someone who has grasped the basics. However, you will need to take into account that even in my time in beekeeping there have been some significant changes, including the introduction of oil seed rape (OSR) that has given many of us a high yield that has to be extracted before the main harvest; the introduction of pests and diseases; the seasons being 2–3 weeks longer; changed legislation and the arrival of the litigation culture.

Don't judge a book by its cover or glossy pictures, but by its content and the experience of the author.

Beekeeping Organisations

In each country there is a national beekeeping association: British Beekeepers Association (BBKA) that largely covers England, Scottish Beekeepers Association (SBA), Welsh Beekeepers Association (WBKA), Ulster Beekeepers' Association (UBKA) and the Federation of Irish Beekeepers' Associations (FIBKA) for the Republic of Ireland. They have either local beekeeping associations below them or county BKAs that are then split into local BKAs. It is these that most beekeepers will belong to for tuition and social events. You can find details of your local BKA on the national association website. Some local BKAs aren't members of a national BKA for a variety of reasons.

To become a beekeeper there is no need to belong to any BKA and there is no registration required as in some other countries. However, with the problems that bees and beekeeping face in the modern day, I believe it is important to join a BKA as you should have access to the latest information, good training and someone to help if you have a problem. Most will include insurance with your subscription and in these litigious days you never know when it might be needed.

Local BKA Meetings

These should be good for learning both practical and theory and are the obvious place to start. A good BKA will have a well equipped teaching apiary with good demonstrators, bee-related winter meetings and a good library. When you first attend I suggest you get to know all the members before getting too involved with anyone, as some BKAs have their resident bores who tend to latch on to beginners and are often difficult to get rid of, especially once they have taken you under their wing. Through talking to everyone you will soon learn who you can rely on and it's often the quiet ones who are the most knowledgeable and helpful, not always the noisy ones. Try to attend all the apiary meetings you can and use your ears, eyes and mouth in the proportion they were provided in.

This will allow you to see colonies at different stages of development and in different weather conditions. Good demonstrators stand out a mile as they will handle a colony gently whilst talking through what they are doing and will be happy to answer questions without making you feel an idiot. They should make calm but deliberate movements and only use smoke when necessary to keep the bees calm, with few in the air.

If your BKA has ability groups, then you are probably put in yours for a reason, depending on your ability and speed of learning or potential to do so. At Wisborough Green we currently have four: Preliminary (for those who have never seen inside a beehive before, or have recently joined and they will stay there until they learn our methods); Beginners (where we would normally expect those with less than a year or two's experience to be); Intermediate (for perhaps another two years); then Advanced Members. We still have some in the Beginners group who have been beekeeping four years, yet some in the Advanced group who have only been with us for a year.

Once you have found someone who is a reliable source of information and help, try to stick to him or her. Don't seek help from several people, as they might think you are playing one off against the other.

Beginners' Courses
These should be tailored to suit beginners, but may vary a lot. Seek out all those within travelling distance, look at the content and don't be frightened to speak to the tutor before enrolling. Some give information that is way beyond what a beginner will need and only results in confusion and an impression of complication. In my experience you can't judge the quality by the price as some may be subsidised or the tutor isn't paid.

Conventions

These tend to be run out of the active season, will probably have visiting lecturers and may be for more advanced beekeepers. Speak to the organisers, tell them your experience and ask if the event would be suitable. There is nothing to be gained from spending a day listening to something you don't understand, but if it is suitable you may well pick up some useful information. Take a notebook and pencil with you.

Magazines

All magazines are of a general nature with none specifically for beginners, as the market simply isn't big enough. Most will have a beginners' section and as the writers are usually appointed their information should be fairly reliable. I still read the beginners' notes and I suspect many other experienced beekeepers do as well. If you don't initially understand an article you can be a bit selective in what you read and return later when you are better equipped to understand it.

Leaflets and Booklets

There are some very good ones available, either free or very reasonably priced. Some will be specialised, but that may be a strength as they should be authoritative. Those from the National Bee Unit (NBU) on diseases are exceptionally good and of course relevant to all beekeepers including beginners.

The Internet

This is a bit like books: there is some very good information, often hidden away in such places as newsletters, but there is some awful drivel as well. In general if it comes from reliable sources you should be OK but some private websites are very unreliable. Being global you will need to be aware that legislation may be different, so check first.

In theory, internet discussion groups should be good and there are some genuine helpful people who are willing to spend their time assisting others, but just as at BKA meetings there are some who are dogmatic and argumentative who can spoil things for others, often made worse when they hide behind a screen and a pseudonym.

2

THE FUNDAMENTALS

Stings

I must deal with this at an early stage. It's a bit like having a warning on a bar of fruit and nut chocolate that says "This product contains nuts"! On more than one occasion I have seen people who come along to a meeting because they want to start beekeeping and ask if we get stung! We now warn everyone who comes to their first meeting: "You will get stung". If you are frightened of being stung then you won't ever make a decent beekeeper.

We often have the noisy clowns who are full of bravery – they aren't frightened and they will be able to take it. A mental note gets taken and often they are the ones who start yelping and leaping around like a gelded dingo when they do get stung. Even though I have seen it many times before it still appeals to my sense of humour, as I know we are unlikely to see them again. I don't need to tell you they are almost always men! I am often impressed with the reaction of many who do get stung for the first time, as very often it's those who make no fuss that subsequently make good beekeepers. They have obviously thought carefully about it and their attitude is right.

Although I never advocate getting stung deliberately it wouldn't be a bad thing if you did several times before your enthusiasm took you too far. In a very small number of cases there is a possibility of a life-threatening situation, although I have never seen one. If there is a likelihood of a problem it's

31

better to know about it sooner than later and for that reason I suggest you are accompanied on your first few inspections until you get stung a couple of times; this can be done at your local BKA.

I do know of one case where a couple had kept bees for several years. The lady suffered anaphylaxis and a couple of years later her husband did as well. They both had a three-year course of desensitisation treatment that initially improved things, then they degenerated to the point where they both had to give up active beekeeping. They did continue with class-room training and helping to run their local BKA which was commendable. I don't mention this to frighten you, but to point out there is a possibility a reaction may come later. However the chances of it happening are extremely small. For the vast majority of people there is no problem at all other than the normal reaction of a bit of pain and some local swelling.

I will not comment on the treatment for stings because I am not qualified to do so. If you do get stung we now know you should get the sting out as quickly as possible, even if you pull it out. Less venom will be injected than if you waste time looking for something to scrape it out with, which is the usual advice.

About Honey Bees
Honey bees (*Apis mellifera*) are the only species of bees out of the nearly 300 in the UK that don't hibernate and thus need to make honey for winter consumption. They are the only bees capable of storing food in any quantity, with the whole colony working hard during the summer to store enough to feed itself during the winter when nothing is coming in. It is some of this store the beekeeper harvests and replaces with sugar syrup. There are over 20 species of bumble bees that, like honey bees are social, but they hibernate and only store food in minute quantities. Throughout this book I will refer to honey bees as "bees" and I don't mean any other species.

Honey bees naturally live in cavities in trees and build their comb from beeswax that is made in the form of scales that are secreted from wax glands on the underside of the abdomen of the worker bees. The bees take the wax scales and form them into the familiar hexagonal cells of the honeycomb.

What Do Bees Bring into the Hive?

There are five main things bees collect and bring into the hive: nectar, pollen, honeydew, propolis and water.

Nectar

Nectar is a sugary substance produced by plants to entice insects to pollinate them, which they do often by accident. Nectar varies considerably and is roughly 20–40 per cent sugar; the rest is water and trace elements. Nectar is gathered by the foraging bees and brought back inside their bodies to the hive, where it is passed to house bees who store it in cells, modify the sugars by enzyme action and drive off water until the water content is around 17–18 per cent. It is then called honey and is the carbohydrate part of the bees' diet.

Pollen

Pollen is high in protein and is necessary for the production of brood food for feeding the larvae. It is collected by the bees and formed into coloured pollen pellets we sometimes see on bees collecting pollen. These are brought back to the hive in what are known as the pollen baskets on their hind legs. The bee backs into a cell, puts her hind legs into it and wriggles herself from side to side to dislodge the two pollen pellets, then turns round and rams them in with her head. Pollen will quickly go mouldy, so the bees fill the cell to within 3–4mm of the top, then, if it needs to be kept for future use, put honey on top and seal it with a wax capping to preserve it for when it is required.

Honeydew

Honeydew is a sugary secretion from insects, such as aphids that suck the sap from trees, and is collected by bees. It is not highly thought of in the UK and Ireland, but is greatly appreciated in some countries such as Germany where bees are migrated to forests to collect it. If our bees gather it, it is included with the rest of the crop where it is largely unnoticed.

Propolis

Propolis is a resinous substance that comes from trees. It is brought back to the hive by the bees in their pollen baskets and used for waterproofing the hive, sealing cracks and many other uses.

Water

Water is needed to keep the humidity of the brood nest correct, for cooling it in hot weather and for diluting honey. That which is driven off the nectar is usually sufficient during a honey flow, but when there is no honey flow they will collect water from puddles, ponds, etc.

What Bees are in a Colony?

Queen

There is normally only one in a colony and she is the only fertile female. She mates on the wing with several drones, normally 10–15, but this varies. Once she commences laying she is incapable of mating again and should naturally live for 3–5 years. She emits pheromones (see Pheromones, page 76) that have a controlling effect on the whole colony; the best known is called "queen substance". During supersedure, when a queen is being replaced, there may be a mother and daughter in a colony at the same time. It is common for there to be more than one virgin queen, but not for very long.

Drones

Drones are the male bees and it was once thought they were lazy and only mated with the queens, but we now know they have a much more beneficial effect on the colony.

Workers

As their name implies they do all the work including feeding larvae, cleaning, comb building, processing food, defence and foraging. They are sterile females and with different feeding at the larval stage could have become queens.

Although there are instances where drones can come from a fertile egg they are usually destroyed by the workers and for the purpose of this book we must assume they are the product of an unfertilised egg. Queens and workers come from fertilised eggs.

Life Cycles

The life cycles are one of what I call the "basics" and should be among the first things learnt as a beekeeper. They are crucial in so many ways. By observation it is handy to be able to tell the ages of larvae as you will constantly need to know them.

	Queen	Worker	Drone
Egg hatches into larvae	3	3	3
Cell sealed	8 (9)	8 (9)	10
Bee emerges	15 (16)	21	24

The above is the time taken in days which may be rounded up or down. There are variations depending on which source you use and I think it is not as precise as is often made out. I believe it makes sense to work on the safe side, especially with the sealing of the cell and emergence in the case of the queen. The figures in brackets are variations that I have found to be quite common, although it depends on what time of the day or night the egg was laid.

Developing Your Technique and Management System

Apart from those who work closely together, or were taught by the same person, all beekeepers will develop their own techniques and colony management methods. In the early days you will probably follow your tutor and for the first few years may well seek his/her advice. When you get more knowledgeable you will either recognise how good your tuition was and continue in the same way, or you will drift off, seek more information and modify your methods. If you have learnt the basics, you will be in a position to understand why other people do things the way they do and you should have gained the knowledge to decide if they are better than what you are doing or not. Try to think things through and if you change I suggest you don't try to mix methods without a lot of thought about compatibility.

As I have already stated I try to keep things simple, but very often I find some methods very complicated and often it appears they haven't been planned too well. Always be prepared to be flexible as no two seasons are the same and neither are two colonies. What you do one year may well be a complete disaster the following year. The trick is to have a simple system that can easily be modified to deal with changing circumstances.

What Ability and Time Do You Need?

I have already indicated that beekeepers need to be much better now than they once were. Due mainly to man's interference, bees have had a pretty raw deal and are suffering badly. Quite frankly they now need our help and understanding and I think they have a right to expect it. There are many people who come into beekeeping and think that all they have to do is have a box of bees at the bottom of the garden, turn on a tap, bottle the honey and they are a beekeeper. Reality is a long way from that, both bees and beekeeping can well do without that approach.

There is learning to do and you need to be fairly fit, with a high level of commitment and understanding. Beekeeping should not be seen in purely commercial terms of getting cheap honey with a minimum of effort. That will result in failure.

You must consider the physical requirements needed and you will sometimes be working on your own. We often have older people who are frail, or have bad backs, come along and say they are going to work with someone else, but sooner or later you will have to do it on your own if the other person is unavailable. If you put an inspection off and you have a swarm you could be filling someone else's chimney up with bees, then you will have the job of trying to remove them. There is heavy lifting to do and if you are frail, things will only get more difficult for you. In round figures a brood box full of honey weighs 55 lb/25 kg, a super 35 lb/16 kg and there could easily be three on a hive.

You would benefit from being practical as hive parts come in flat packs and need nailing together and there is maintenance as well. Eyesight should be good as you will need to see small eggs.

Although the basics must and can be learnt quite quickly, you will need to learn enough to keep bees healthy and alive. That means learning about diseases and their treatment. It's not simply a case of, "Oh I think my bees have got X so I'll give them a dollop of Y"; you will need to understand the pest or disease, give your own diagnosis and treat accordingly. Bees are not like plants, they are animals.

In short anyone who is active and capable of quick learning can keep bees. If your garden is small there are usually other people who will let you put bees on their land in exchange for a few jars of honey a year. There is a reasonable level of commitment needed and that is around an hour a week throughout the summer per colony for a beginner, but after perhaps a year or so that can be drastically reduced when you learn to do things quicker. If you attend BKA meetings that might involve something like four hours every fortnight during the summer, then there is time helping others as we all do. Beekeeping can be time consuming.

How Do You Start?

I am in absolutely no doubt the first move should be to contact your local BKA, where you should be welcomed. All BKAs are run by volunteers and their character will always reflect the membership and key officers at the time. They do vary a lot and my advice would be to visit all of those within a reasonable distance that you are willing to travel to on a regular basis, bearing in mind it could be a dozen times or more a year. Check them all out with particular attention to the level of tuition, attitude, friendliness, number of meetings, facilities, winter meeting programme and, most importantly, if they have a well organised teaching apiary. If your time is limited make sure their meetings fit in with your work or social life.

Have a look at their websites and winter programmes and make sure they are bee-related; you join a BKA to learn about beekeeping not penguins or vintage fire engines.

When you have made your decision please treat them as your BKA and show loyalty by attending meetings and helping where you are able. Each BKA needs people with non-beekeeping skills in order to run efficiently and you may be the only accountant or carpenter they have. If you are happy to drive 20 miles to take advantage of their benefits, you shouldn't use distance as an excuse not to help.

A BKA should support its members if they need help and when you have gained experience perhaps you can help someone else. One important benefit could be insurance, but you will need to check to see if that is offered.

I encourage all our newcomers to handle bees a lot before they get their own. Very often initial enthusiasm dies after a couple of stings and all of a sudden we hear that the person is "allergic". They rarely have a problem; they simply don't like getting stung and it's much better to find out about fear before any bees or equipment is sourced. In some areas there is a high turnover of beekeepers in their first three years and I think this is because many buy bees without much guidance and before they are ready or know what is involved.

I will deal with the purchase of bees and equipment later, but you need to know there are different kinds of both. It's important, less troublesome and less costly if you get it right first time. I tell beginners that starting beekeeping isn't a simple case of buying any old hive of bees and that's the end of it. You will need to know what hive and equipment to buy and where you can get bees. I usually give all this information, but warn people of the dangers of buying anything without help.

It is difficult advising on how to start as there are so many ways and personal circumstances vary. If someone is giving you a present then try to make sure it is something you need or perhaps a voucher. You may be in the lucky position where someone is giving up or moving away and you are being given the bees and equipment, in which case you may not have any option other than to start with what is on offer. If it's different from what you are used to, stick with it and see how you get on; you can always change later.

If you have an opportunity to acquire second-hand then seek sound advice before agreeing a price or to buy. There is always a possibility of disease and although most cases I have dealt with have been trouble free, it is much better to take advantage of the experience available. If the vendor is known, or the bees have been checked and are free from disease, it is probably safe to use any spare equipment after cleaning it, but if they are unknown it would make sense to scorch out hives and to burn any old frames and combs, especially if there are no live bees. Always be prepared to walk away from potential trouble.

If you buy second-hand, beware of any home-made hives. I have been the auctioneer at West Sussex BKA annual auction for many years and have sold many hives. Some home-made ones are very well made, others aren't, so take a tape measure with you and a very keen eye. A guide price for home-made equipment, even if well made would be around a quarter of the price of the equivalent second-hand machine-made item in the same condition.

If you are buying everything new you have the option of buying the whole lot, or piecemeal. I would suggest the latter, as I think it better to start gradually. I have seen some "Beginners' Packages" and advise against them as there are often bits you don't need or of poor quality. You may pay a bit extra, but most things will last a lifetime if looked after well. Beekeeping is a hobby and the enjoyment will be lost if you are struggling with poor kit.

I must emphasise caution. I have seen so many people charge headlong into beekeeping, only to find it doesn't suit them or, more commonly, they don't like being stung, then give up quickly after spending a small fortune. Get your grounding on someone else's bees, have a chat to a demonstrator, ask for an honest opinion on how you are getting on and if fine then think about buying your own. That in my view is one of the most important functions of a local BKA. If you do decide to proceed I suggest you have two colonies fairly quickly and you should have gained the knowledge and experience to do so. Colony losses and queen failure are much greater than they once were and if you have a problem with one colony you have the other to help it out.

3

THE BEEHIVE

Man kept bees in primitive hives for several thousand years until it was discovered they always leave a number of passageways of about 6–10mm wide so they have easy access around their nest. This gap has become known as "bee space" and the bees rarely fill it, allowing us to make hives in box form and put comb in frames with a bee space between box and frames. This means the beekeeper can easily take the frames out of the box and inspect the combs. This discovery was only fully made use of in the mid nineteenth century. Great strides in the design of beehives were made in a very short time, but have hardly changed since, with the hives available today using very similar frames.

Parts of the Beehive
A modern beehive (see Fig. 1 overleaf) is made up of the following parts from the bottom:

Floor – Fig. 1(f)
This until recently was solid wood, but now it is normal to have a framework and wire mesh over the top of it, which is called an Open Mesh Floor (OMF) and is used as an aid in the control of varroa.

Brood Box – Fig. 1(e)
A topless and bottomless box containing combs where the

The Beehive

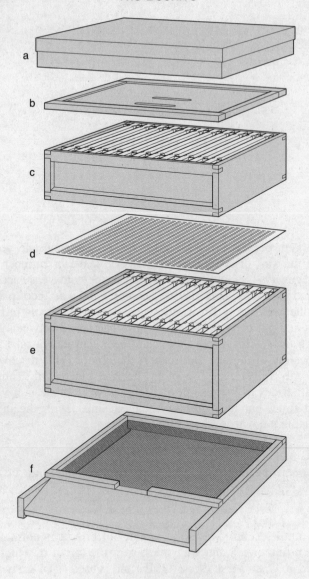

Fig. 1 Parts of the beehive

(a) Roof

(b) Crown board

(c) Super

(d) Queen excluder

(e) Brood box

(f) Open mesh floor

brood is reared and the bees spend the winter. Usually immediately above the floor. Some beekeepers use two if their bees are prolific and need extra space. This is called "Double brood box". Brood boxes can also be used as supers.

Queen Excluder – Fig. 1(d)

Made of metal or plastic with slots that restrict the queen to the brood box so she can't lay eggs in the supers. The slots are of a size that only workers can get through, not the larger queen and drones.

Supers – Fig. 1(c)

These are topless and bottomless boxes of the same design as the brood box, but shallower and there can be several on a hive depending on the needs of the colony. Most of the honey is stored in combs in the supers and harvested by the beekeeper. The term "super" is used either with or without frames. If a super is used as an addition to a brood box and is used for brood rearing it is termed "Brood and a half".

Crown Board – Fig. 1(b)

A board, usually made of plywood, with wooden strips around the edges, top and bottom to create a bee space. There are usually one or two slots or holes allowing bees to access a feeder that is placed on top, or to be fitted with Porter bee escapes (see Clearing Bees From Supers, page 121) when it is referred to as a clearer board that is used for clearing bees from supers before honey harvest.

When used as a crown board I always cover the holes in the summer. If you don't, the bees will be able to come up through them and build wild comb in the gap between the crown board and the roof, making it difficult to remove the roof. I use bits of thin flat wood to cover the holes not Porter escapes otherwise the bees will propolise them up.

I always have the slots running at 90° to the frames throughout the year as a matter of course, so that during the winter when they are open I can see how much food is left (see Chapter 11, Wintering).

Roof – Fig. 1(a)
Usually ventilated and metal covered. The depth of the roof doesn't matter, but deep roofs are considered helpful in very exposed areas.

Assembling Hives

All new hives should now be made so they are fully interchangeable, whoever makes them. The premium grade ones are made from Western Red Cedar, are usually knot free and are obviously the most expensive. The quality is excellent and better than some furniture you can buy. You have the option of ready assembled, or the more common flat pack for self-assembly, with a slight price reduction.

Some of the major manufacturers now supply seconds that are either the premium grade with knots in or are made from a lesser grade of timber; these are typically half premium price or less. I have been involved in buying many of these seconds for my local BKA and its members and, provided you sort them out, so you don't get knots where you need to put a nail, I think they are very good value and if looked after will last a lifetime. They tend only to be available at certain times, but in my opinion are worth waiting for. They should come with nails and you may need to look in a catalogue or on a website for assembly instructions.

Plate 2 shows a flat-pack ready for self-assembly. You can buy whatever you like but for a beginner I would recommend as a minimum a floor, brood box, three supers, crown board, queen excluder and roof. You can see that these are seconds as they have a number of knots in them. If selecting yourself at a sale, try to avoid those with dead knots in case they fall out.

It might appear simple to assemble hive parts, but many people have difficulty and have to take them apart again. It would make sense to have guidance.

Some people advocate gluing and nailing hives, but I have never glued and I would advise against it. A hive is a living thing and moves due to weather and use. When it does move, it puts pressure on the small area that is glued, where if it was only nailed there would be some "give". Once glued it is almost impossible to take apart without damaging the wood, so if you get it wrong, as is often the case even with experienced beekeepers, you can't get it apart again and this is the same if you need to make a repair. I know of hives that are in excess of 50 years old that have only been nailed and are as good as the day they were made. When assembling a National brood box or super I suggest you run a plane along the exposed underside edge of the top bar so it's more comfortable when lifting and carrying hives.

The nails that are usually supplied are what are called "lost head" nails and have small heads. They are perfectly adequate, but if you need to take the item apart to make a repair I find that the nail doesn't always pull out, but stays behind and you end up pulling the head through the wood. I use what are called "wire" nails that have a bigger head that prevents this problem. When nailing, hit the pointed end of the nail with a hammer to blunt it so it doesn't split the wood. In cases where a nail must be driven through a knot, drill through the knot first.

There are alternatives to cedar such as softwood that are offered by some manufacturers. This will be quite adequate provided that the parts that are outside all year are treated with preservative, but supers should last a long time without any.

From time to time there are companies who start up offering cheap hives. Some of them make good quality equipment, but some don't. As always have a look before buying.

Assembling Frames

When making frames they need to be flat, not twisted and I do this on a flat surface so I can tell if they rock. There are several ways of holding frames together and I prefer to use gimp pins that are sold by equipment suppliers. My method of frame making is as follows:

• At the bottom end of the sidebar there is often a sliver of wood if the groove is cut off centre. Go through the whole pack and cut these out with a sharp knife, otherwise the foundation (see Comb and Wax Foundation, page 49) will get held up when you slide it in. See Fig. 2(a).

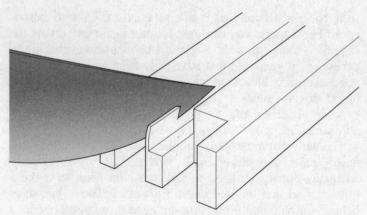

Fig. 2a Cutting out sliver of wood from the side bar

- Don't cut the wedge out of the top bar yet as you may cut yourself; leave it in.
- Put both side bars in the top bar with the grooves on the inside. See Fig. 2(b).

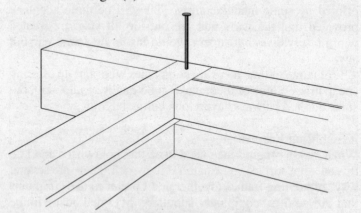

Fig. 2b Side bar inserted in top bar with groove on inside

- Knock the sidebars home, making sure they aren't twisted by lying on the flat surface.
- Put one pin in each side bar to fix it to the top bar.
- Put both bottom bars in and tap with a hammer to get the ends level.

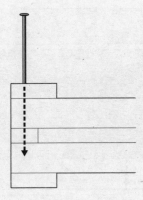

Fig. 2c Pin through side bars and bottom bars but not through groove

- Drive a pin in each from the side, but not through the slot as this will stop the foundation sliding in the groove. See Fig. 2(c).

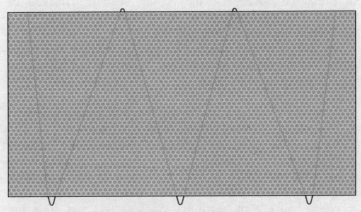

Fig. 2d Sheet of foundation

Some people drive one pin through each bottom bar into the side bar from the bottom so they can be removed later for cleaning. Bees build comb between the frames, fixing the bottom bar of one frame to the top bar of a lower frame and often when I have taken a frame out of a box one of the bottom bars has been left behind. They are then difficult to

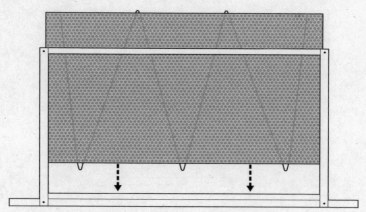

Fig. 2e   Sliding the foundation in

replace, But with my method this doesn't happen.

If you don't need the frames yet, leave the foundation out otherwise it will go stale. Store the frames in an empty super or brood box otherwise they could become damaged or twisted. Don't put a piece of string through them and hang them up in the shed as they will distort.

Fig. 2(d) shows the sheet of foundation with the long loops at the bottom, short loops at the top. When you need to put the foundation in the frames:

- Remove the wedge from the top bar with a sharp knife.
- Trim off any slivers from the wedge or frame.
- Separate the bottom bars and hold a finger on the inside so your finger gets pinched.
- Slide the foundation in from the bottom of the frame with the longer loops leading. See Fig. 2(e).
- When halfway in bend the loops at 90° and slide the foundation home.
- Put the wedge back and secure with two pins.
- If there are loops sticking through the bottom bars then pinch the bars with your fingers next to the loop and force the loop between the bottom bars with a hive tool. This will prevent the uncapping knife being snagged when you clean up the frame during uncapping.

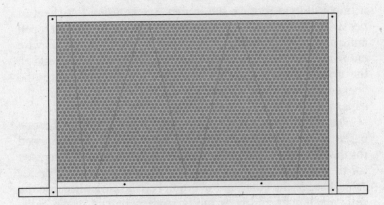

Fig. 2f Finished frame

This should use six pins per frame and it will be as strong as you can get it. See Fig. 2(f).

Comb and Wax Foundation

Natural comb is made from beeswax by the bees and has a midrib, with the familiar hexagonal cells horizontal, but pointing slightly upwards on each side when you are viewing from the end of the comb. The cells are normally in two sizes, smaller for workers and larger for drones. They are multi functional and are used for rearing brood as well as storing honey and pollen. Queen cells aren't normally horizontal, but vertical and about the size and shape of acorns. Worker and drone cells are constantly reused but queen cells are only used once. The bees place the honey at the top and to the side of the combs, with brood below. Pollen is stored between the honey and brood, usually only above the brood with imported bees, but all round and often in between the brood by native bees.

In a natural nest the brood area is forced downwards as the honey store increases during the spring and summer, but moves upwards during the winter as the bees use their stores. Both brood and honey are sealed with a capping, the brood to pupate and the honey to preserve it.

If bees were allowed to make a nest in a hive with empty frames they would build comb wherever they wished, making it impossible to remove the frames, so we use a flat sheet of beeswax that is embossed with the pattern of the base of a cell and fix this in the frame. It is called foundation and has wire embedded in it to strengthen the comb. The bees build cells on the foundation and make nice flat combs, allowing the beekeeper to remove the brood frames for inspection and the super frames for honey extraction.

Foundation is available in two cell sizes, worker and drone based. Apart from the odd occasion, worker base is usually used for the brood box, but either worker or drone base can be used for the supers. It is a matter of personal preference. I prefer drone base as the honey extracts quicker and easier because the cells are bigger and bees rarely store pollen in drone cells, making uncapping easier and avoiding wasting the effort of the bees in collecting and storing it. Some people will tell you that if a queen gets through a queen excluder you could have a lot of drones in the supers, but there are only three ways a queen will get into supers, because she is small and poorly raised, the queen excluder is damaged, or due to careless colony manipulation, all of which can be easily avoided.

You may have read about the "warm" and "cold" way of arranging a hive and wondered what this means. It is simply the orientation of the frames relative to the entrance. If the brood box is placed on the floor with the frames running perpendicular to the entrance, it is thought that draught will go in the entrance and up between the frames, so this is called the "cold" way (shown in the upper part of Fig. 3); if the frames are parallel to the entrance, the draught won't go between the frames, hence this is the "warm" way. In fact, it makes absolutely no difference, especially now OMFs are widely used. This is one of beekeeping's many myths, although it is useful terminology to describe how a colony is set up.

In most winters I don't think it makes much difference but I do have a preference for the cold way, because in a prolonged cold snap when bees are clustering tightly they may run out of available food and won't break cluster to go over, round or under the last frame to get onto fresh food. On many occasions I have seen a colony that was set up the warm way starve with plenty of food they couldn't access. This is called

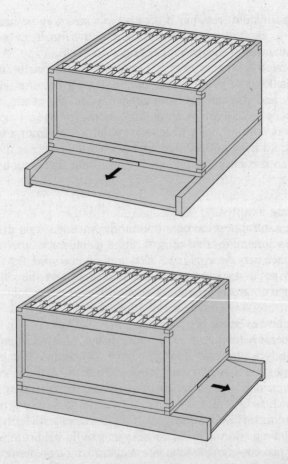

Fig. 3 Arranging a hive: cold way (upper) and warm way (lower)

isolation starvation and is a depressing sight, especially for a beginner. Don't change from warm to cold way or vice versa after the end of August, as the bees will have arranged the brood nest so they know where everything is.

It is easier to manipulate a colony with the frames running left to right and if you are inspecting all the frames it is awkward to lean over the whole brood box, so I find it easier to work from both sides. This is easy when your colonies are set up the cold way, but if set up the warm way you need to stand in front of the entrance where you cover it up and end up with a lot of distressed bees on your back.

Interestingly it is only a couple of hives, including the National where you have the option, simply because they are square and you can put them either way, all the others are the cold way unless you make modifications. In the vast majority of cases where I have removed a wild colony from a tree or building I have noticed that the bees build their nests the cold way, so the bees are telling us something and we ought to listen.

Spacing Frames

In the natural nest the bees build the combs a certain distance apart, allowing for two cells of brood facing each other with a bee space in between. I have measured several wild nests when removing them from trees and buildings and the distance between combs is usually 35–38mm. This is dependent on the worker brood where the length of the cells is usually about the same; however, the bees can make the cells that contain honey considerably longer, allowing more distance between centres if we space combs wider apart. We have standardised the spacing of combs and have two sizes, narrow (36.5mm) in the brood box to correspond with the length of a cell containing worker brood and wide (47.6mm) in the supers, although these can vary a bit depending on the manufacturer. If super frames with foundation were put on wide spacing, the bees may build wild comb in the wider gap between the foundation instead of on the foundation, so we usually use narrow spacing until they are drawn out, then use wide.

There are several ways of spacing frames, but it is usually achieved in one of three ways, allowing you to get 11 frames in a National box on narrow spacing or 9 on wide:

Metal or Plastic "Ends"

These slip over the lugs of the frames. I find them a nuisance as the bees tend to propolise them up, making it difficult to remove them as you have to when fitting them in some honey extractors.

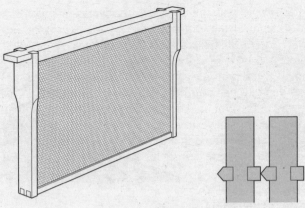

Fig. 4 Hoffman frame

Hoffman

Hoffman spacers where the spacing is achieved by the side bar of the frame being wider, so making them self-spacing. See Fig. 4. The standard width is 35mm, allowing you to get 12 frames in a National box, but I advise against that as it is possible you will damage bees when removing the first frame. A division board is normally used which creates room when it is removed.

I don't like Hoffmans as they get propolised, you can't get them closer together if you need to, you need some other sort of spacing if you want to use wide in the super so effectively they are wasted and they use more wood. They are very popular, but I think only because few know of the benefits of castellated spacers.

Castellated Spacers

Castellated spacers (see Fig. 5) are simple pieces of sheet metal with slots in that are nailed to the boxes. Although not perfect, in my opinion they are the best form of spacing by some distance for both brood boxes and supers. Owing to there only being point contact with the frame they don't get heavily propolised and in operation you can lift the frames out of the spacers and close them up enough to make room to get frames out of the brood box without "rolling" bees. They are

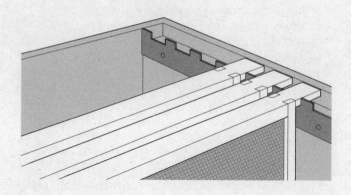

Fig. 5   Castellated spacers

much criticised by people who don't know how to use them properly, if indeed they have used them at all. You can use the cheap standard frames. There are 9-, 10- and 11-slot available and some people use 10 slots in all their supers as bees will usually build out foundation without building wild comb in between. Most of my supers have wide spacing with a few narrow for drawing out foundation and these can also be used for cut comb honey.

Care of Wooden Beehives

I won't deal with the materials used in the preservation of beehives, otherwise it might date this book due to changes in legislation, availability of new products or withdrawal of old ones. Instead I suggest you consult other information sources such as advisory leaflets or manufacturers' websites. I know a bee farming family who have some National hives well in excess of 50 years old and still in very good condition considering they get far more use than most. They have looked after them by keeping them dry and treating them regularly and there is no reason why they won't last another 50 years. I have seen far more hives that have rotted than have been worn out, which I think is a great pity. One major problem is damp and if kept off the ground with vegetation trimmed away the life will be extended considerably.

It is unwise to paint single-walled hives with gloss paint. The moisture that is created inside seeps through the wood and is unable to disperse because the wood doesn't breathe. As well as making the combs damp, it often causes the paint to blister and peel off.

Although I have never done it myself I know beekeepers who coat their hives with either linseed oil or rapeseed oil and they are apparently good, presumably the oil is acting as a water repellent.

Beehives – How Do You Choose?
There are about ten types that are easily available and you will find it an expensive and troublesome exercise to change once you have started, especially if you have bought new and you can't sell because few other people use it. If you have inherited bees you may not have an option, in which case it would be sensible to stick with what you have until you have gained enough experience to change, if that is what you want to do.

There is more twaddle spoken about beehives than almost any other beekeeping subject and often by people who couldn't recognise all the types available. You will often be advised on which hive is best, without being told there are several issues involved. Some kinds of bees are much more prolific than others and the queen needs a far larger brood box to lay eggs in than less prolific queens do. Many beekeepers simply don't understand that the bees must suit the hive. It is unwise to try to keep non-prolific bees in a large brood box as they will fill the area the queen doesn't need with honey. Conversely, if you keep prolific bees in a small brood box you could end up with the colony swarming more readily and it won't be able to store enough food to survive the winter.

The need for more space by prolific queens is usually overcome by either using larger hives that can be very heavy, or two boxes of a smaller hive which can be another brood box with frames of the same depth and referred to as "double brood box", or a super that is shallower and referred to as "brood and a half". In my opinion using two brood boxes is very unsatisfactory as you have twice the number of combs, making it much more difficult to manipulate and to find the queen; and with brood and a half you have brood on two sizes of frames, the latter being one of the most annoying things I have found in beekeeping.

Confusingly there are several frame sizes, but we have the British Standard frames that are the most popular by some distance. They are the only ones that have long lugs and having had experience of all types I find that a great benefit. They come in three depths: "shallow" for supers, "brood" for standard brood boxes and "deep" that are also known as 14 x 12 for extra deep brood boxes. The latter have gained increased following in recent years, but I think they are unnecessary as they are certainly too big for my bees. Several beekeepers I know have tried them thinking they would be better, but many have rejected them as they have found them too big and heavy to handle and too big for non-prolific queens. They will not fit in most honey extractors which will reduce flexibility. I have been told on several occasions that the combs can fall out of the frames when they are handled in hot weather, due presumably to their size and weight. I would strongly advise beginners not to use them until they have enough knowledge and experience.

I won't deal with all the different hive types, but will concentrate on those that take the British Standard (BS) frame that is used by the vast majority of beekeepers in Britain, so it is easy to buy and sell bees and equipment.

The ones I suggest you consider are:

WBC

WBC is the one that is instantly recognised as a beehive with the sloping sides and often painted white. It is the only double-walled hive and consequently there are a lot of parts, making it expensive and cumbersome to use. At one stage it was the most popular hive, but that is no longer the case. It needs a lot more maintenance than other types. If you just want to stick to a couple of hives the WBC is suitable, but if you wish to expand from that I advise a single-walled hive.

National
This is the most popular hive by some distance; at Wisborough Green well over 80 per cent of hives in use by members are Nationals. Its construction makes the boxes easy to lift, it is light and as it is popular is available second-hand and easy to sell. The brood box is the right size to accommodate non-prolific bees without them running short of food in the winter, although prolific bees may want an extra box. I thoroughly recommend this hive as I think it's suitable for our conditions and it has been my main hive type for many years.

There are two versions of the brood box and super of the National although they are totally compatible. All that varies is the method of construction. The original had straight sides with handholds machined into the faces; the later version that became known as the Modified National is more economical on timber and has got a rail that creates a much better handhold. I only mention this because although they haven't been available new for many years there are still a lot of the old pattern parts in use and the beginner may not know what they are. They are now all known simply as National.

Smith
Smith uses the same size frames as the National and WBC, but with short lugs. It has a simple method of construction and is easy to make if you have the woodworking equipment. There are handholds cut in the side boards and as these are only shallow the whole weight of the box is on your fingertips when lifting, although this can be overcome by screwing plinths on the sides. It is very popular in Scotland, so potential beekeepers there should seek local guidance before purchasing.

Other Hives
I suggest you stick to one hive type or at least one frame size when you have a small number of colonies. It makes everything interchangeable and overall you will need far less equipment, as well as keeping you good-tempered.

Polystyrene Hives
Polystyrene hives have been used successfully in Europe and especially in Scandinavia for a long time. They are becoming widespread here especially with commercial beekeepers, because they are lighter and cheaper than wood. I have never owned any myself, but have handled bees in them on many occasions. At the time of writing I have reservations, especially for beginners. The vast majority are Langstroth (American design) that don't use BS frames, although there are now some National, but most of those I have seen aren't compatible with wooden hives, so you may not be able to use both together on the same hive. The polystyrene must be high density otherwise they are easily damaged, bees will chew them and they won't last very long. Although some are clearly good quality and high density, others aren't. Owing to environmental issues disposal of damaged hives may be a problem and I don't want to get into the argument about whether polystyrene or wood is more environmentally friendly! You would need to satisfy yourself about these and any other issues before buying.

Alternative Beekeeping
I must include the use of the top bar hive (TBH) and the Warré hive as they are now both firmly part of beekeeping, albeit fairly small. I have no experience of using the Warré and only a little of TBHs where I have handled bees on several occasions. For that reason I think it unreasonable for me to make much comment about the actual methods and further information should be sourced from elsewhere.

None of the methods uses frames in the same way as conventional hives and they use little or no foundation, allowing bees to build their combs how they wish as they would in the wild, or as they did in skeps (straw or wicker beehives traditionally used by beekeepers). I find that the TBH is fairly easy to manipulate, but because there are no frames the technique of handling the combs is different, otherwise they are likely to part company from the top bars to which they are fixed.

The term used for this type of beekeeping is "Sustainable"

or "Natural" Beekeeping and from what I have seen and read there seems to be considerable differences of opinion about the methods, their effectiveness and what is natural and what isn't, with as many disagreements as there are with conventional beekeeping. Among the differences are feeding sugar, using soft chemicals such as thymol or organic acids to help in the control of varroa and whether the hive design will allow the bees to build a vertical or horizontal nest.

Sustainable beekeeping has become very emotive with extremist views in all camps. There are accusations that all conventional beekeepers exploit bees entirely for their own gain and everyone treats with chemicals in order to gain maximum honey crops. There will always be those who do, but they are in a very small minority. The vast majority of conventional beekeepers I come into contact with have the welfare of bees very high on their list of priorities and I simply don't see everyone as being the sort of beekeepers that are often portrayed. I come across many beekeepers who find extracting and honey handling a chore and quite often leave the honey for the bees. They are more interested in their bees and keeping them healthy. I have come across many people who have made equally adverse comments about sustainable beekeeping who have clearly had no experience of it, yet are very vocal in their condemnation. Both extremes are unfair in my opinion as I don't think a balanced view will be given without a reasonable amount of experience with both methods and a full understanding of what each one is trying to achieve. There are unfortunately many similar instances in beekeeping and the beginner should always be aware of prejudice and intolerance.

My position is that I don't care how people keep bees provided that they look after them well and avoid a nuisance to other people. I accept that in teaching beekeeping I largely use one method that has been developed over a long period with experience of many other methods. I have tolerance of other views provided that the thinking is sound.

I find it highly commendable that some are concerned about the use of chemicals and the welfare of bees as I do, but in the modern day when we have to deal with "modern" problems I think it is irresponsible to return to what is effectively "let alone" beekeeping, where colonies are left to fend for themselves and swarm in an uncontrolled way, without any concern about the possibility of filling up someone else's chimney with bees they don't want.

Over the years I have removed many wild colonies from trees and buildings and until the arrival of varroa they were usually healthy, but now they are often very weak and riddled with varroa and its vectored viruses, especially deformed wing virus (DWV) where the bees are unable to fly (see Varroa, page 135). The same thing happens in a conventional hive when it gets abandoned, so the logic in me suggests that not inspecting colonies on a regular basis, as often advocated, will not result in anything different and I don't see the methods of sustainable beekeeping advised by some as a way of treating bees better, which was the aim in the first place.

From what I can understand a lot of sustainable bee-keepers have much higher colony losses than conventional beekeepers, although I accept that pre varroa their losses may well have been lower. With conventional beekeeping there is a chance of addressing disease with non-chemical methods, but with some sustainable hives it seems almost impossible to inspect the colony to check for disease. There are currently four notifiable pests and diseases that you need to look for and deal with if you see them. Whatever hive you have the Bee Inspector needs to inspect it easily.

I have come across people interested in starting beekeeping who have believed what they have been told and read that sustainable beekeeping will solve all the problems bees face. Beginners must make sure they are not drawn into something simply because of what is effectively propaganda and may be untrue. There are some fairly wild claims made by a few vociferous people and I suspect they annoy the more respon-sible members of the sustainable movement.

I have an open mind on sustainable beekeeping and in some ways I myself find it attractive. Very little equipment is needed and hives can be easily made from recycled material, making beekeeping incredibly cheap, with no standard sizes, so they can be made whatever size you like.

Beekeepers now have to be much more knowledgeable than in the past and I strongly suggest learning conventional methods before going down this road. That way you will see if the objections to conventional beekeeping are correct and you will learn about the workings of a colony.

Siting Hives

Bees naturally live in trees, usually several feet off the ground, but we tend to keep them close to the ground which introduces several problems that we need to overcome. Dampness can be addressed by making sure the site is airy, without having a wind tunnel effect such as can be caused by a tallish gap between buildings, solid fence or dense hedge. Long grass around the hive is beneficial in the summer, but should be cut away for the winter. Full sun for long periods in the summer could cause the bees problems with ventilation, although it isn't so important if you use OMFs.

As most people have gardens this is usually the first place they think of to put their bees. Many gardens are suitable if a bit of common sense, understanding and tolerance is used. If you have near neighbours it would make sense to tell them you are going to keep bees and ask how they feel about it. Expect them to be concerned if they have young children or a genuine fear of bees, as many do. "My wife is allergic" can mean the husband is frightened, so be careful!

Plate 3 shows how bees and children can live together. That particular garden also includes chicken and a sizeable vegetable patch. The three hives have a solid fence behind and trellis in front. The bees probably won't fly through the trellis unless their entrance is very close so will immediately be forced up above head height, allowing the family to enjoy their garden. If there were quick growing climbers, e.g. honeysuckle or clematis, planted by the trellis, the bees would soon be hidden.

It makes sense to keep bees a reasonable distance from the house and if you have somewhere like an orchard or "behind a shed" that could be ideal. Fine plastic netting can be used to force the bees to fly upwards.

If you have an allotment that may be a possibility, but I

think there are dangers and if there is a possible alternative I would try that first. I have never had to keep my bees on an allotment, but I have known beekeepers who have and there seem to be more problems than if bees are kept elsewhere. The owners or trustees would obviously need to be consulted and they will put the interests of non-beekeepers first. There may be rules already in place. It seems that vandalism is quite high on some allotments and this may need dealing with. To help those with little or no alternative I think the more bees that are kept on allotments overall the better, but I think it reasonable that the standard of beekeeping should be high to prevent swarming and to avoid bees becoming bad-tempered due to bad handling.

In the past, most bees were kept in the country where neighbours were genuine country people who knew that bees were an important part of country life. This is no longer the case as many people are paranoid about being stung, probably fuelled by stories of "killer bees".

If your neighbours are unhappy about your keeping bees, there is no point in falling out with them, as sooner or later you will have a problem anyway. If they get stung by anything, even if it isn't a bee, you and your bees may get the blame. I think the sensible thing to do is to find somewhere else to keep them.

Most people will be able to find a piece of rough ground fairly close to home that can be used in return for a sweet rental, but make sure you aren't likely to cause a nuisance to someone else. The going rate is normally a jar of honey per year per hive.

I encourage all beekeepers to be good and responsible, as making regular inspections and knowing what you are doing is important in encouraging non-beekeepers to share our enthusiasm. If there is no problem they can't complain.

Wherever you have your bees, they should not be a nuisance to anybody else and much more thought is needed if your apiary is near habitation or a right of way. Don't create a possible nuisance by placing a hive near a window, path or next to your neighbour's property unless there is a bee-proof barrier in between. Despite your best efforts you will have swarms and these are just as likely to cluster on your neighbour's property as yours.

If bees are foraging in front of their hive they could fly at around head height for some distance if there are no obstacles. To avoid this you can force them up to a higher level by facing them towards an obstacle such as a hedge, solid fence or a building. Keep them away from washing lines otherwise the washing will be covered in faecal matter, especially in the spring. Bees are likely to fly in any direction at any time depending on where their source of forage is.

When selecting a site on someone else's property you need to think reasonably long term and get it right to avoid the inconvenience of moving. Easy access is useful, by vehicle, wheelbarrow or carrying distance. Full supers weigh a bit more than empty ones, or they do after you have carried them 100 yards, which includes negotiating a barbed wire fence, stile, several rabbit holes and a ditch full of stagnant water. Be aware that if your apiary is easily accessible for you, it is for someone else as well and bee rustling has always been an occasional problem. It may pay you to mark your hives in some way; I think branding is favourite as it's easily visible. It doesn't have to be an expensive brand, as you could soon make something up from a piece of strip metal. Brand a spare piece of wood and take photographs of hives to help with identification if needed.

During a break in writing this book I spent a day walking with three dogs and a beekeeping neighbour. She spotted 12 hives that I had just walked past without noticing that belonged to a commercial beekeeping family I have known for nearly 50 years. They were close to a footpath and 30 yards from a road we have both travelled along for many years. We didn't know they were there, although we knew there were some bees in the area and had often looked but never found them. This shows the value of seeking sites that are hidden.

Your hives must be secure against animals, so a fence may need to be erected with a gate. There should be plenty of room to work, so leave space between or behind hives to put the roof and supers you take off each hive. Keep hedges and bushes trimmed as they can ruin your veil.

The direction hives face doesn't matter at all as it's the temperature of the hive that is important not the aspect. I face my hives in all directions and they all perform the same. Unless you have a lot of hives in windy, open and featureless ground it doesn't matter if you keep them in a row.

Hives should be set up level with the top of the brood box at a convenient height to inspect the bees comfortably, bearing in mind their height and you need to allow for at least three supers. Too high and it might be difficult to remove the top super; too low and you will soon have backache inspecting the brood box.

The stands should be sturdy, stable and not likely to collapse, as each hive can easily weigh 150 lb in a good summer. My stands take two hives and are two concrete building blocks stood on their sides, with a stake each side of them to prevent them falling over "windscreen-wiper fashion". On top of that are old timbers that should be at least 50mm x 50mm x about 1.5m long and braced together with two other pieces of wood. These are cheap as the materials can often be found in skips. If there is plenty of room, hives can be placed singly, but if space is limited you can put hives in pairs and use the space between the pairs to stand when inspecting hives.

I am often asked how many colonies you can have in one place. It's difficult to answer as it relies on so many things, including the amount of nectar and pollen-bearing plants available, how many other colonies there are in a three mile radius and the prolificacy of the bees. However, in general you can have at least a dozen in one place and that is enough for a beginner.

In reality, there are very few places that are unsuitable for bees, although some are better than others. I like fairly open woodland, as shown in Plate 4, as it is usually airy with dappled shade and bees can fly out of holes in the canopy without causing problems to anyone. The hives illustrated are in an ideal spot and are unlikely to be in full sun for a lengthy time. The site is fairly open and airy so unlikely to be damp in the winter. The hives are on stands in pairs and set out so there is room to handle the colonies without being cramped.

In looking for a site you should make sure you have good access and that animals can't knock the hives over. Water is needed by bees, but I have never made provision for it as most areas have a natural source close by. If you do wish to provide it there are many ways of doing it and I'm sure you can find one, but give them something to land on to prevent them from drowning. It will certainly provide interest and is usually a good way of finding out if there is a nectar flow on as bees will desert it. However, when siting your hives, check that there aren't any nearby streams or rivers that could lead to flooding. Also ideally the hives should be away from a road or footpath.

If hives are placed in a line close together in an exposed, featureless and windy place the bees may fly into the wrong hives having been blown by the wind. This is called drifting and is not normally a problem for a beekeeper with a few hives that have a bit of shelter. If there is likely to be a problem it can usually be overcome by leaving a reasonable distance between the hives and facing the entrances in different directions.

Lastly can I suggest you stop reading now and move right away from all other family members? OK? Now carry on reading, but don't show any emotion whatsoever. Do you remember those two hives I suggested you have, so one can help out the other in case of any problems? That will sound reasonable to anyone and can easily be explained. Occasionally you will need room for more, as some operations may need it, especially during the summer and this is a normal part of beekeeping. That can also be fairly easily justified. What is very common is that many people actually like beekeeping and the extras become permanent. That may be a bit more difficult to justify. Shhhhhhh!!! Got the message? Now move back to where you were, carry on reading but still show no emotion. It's called bee fever!

4

BEES

Many people including beekeepers think that all bees are the same, but that is definitely not the case. There are probably more beekeeping arguments about the type of bees that are best than anything else, usually by those who have little or no experience of all of them in any numbers. I have known many people who have bought a queen of a particular race "just to try them" and base their opinion for the rest of their beekeeping existence on that one colony. It really isn't as simple as that because all the sub-species have different characteristics and may need different management techniques from others and there is often considerable variation within the same race. There is no point in making judgement based on a small number of queens if you have no understanding of their needs. It's a bit like buying a border collie puppy and thinking it will behave the same as every other border collie and need the same treatment.

It is now widely accepted that none of the sub-species that are likely to be kept in the UK and Republic of Ireland is totally pure, as they have been mongrelised by accident or design, but for the purposes of this book if I refer to "pure" I mean in the purest form available. I have a reasonable amount of experience of all sub-species and I have never experienced genuine bad temper in any of them when pure, despite much writing to the contrary (especially concerning our native bees), some of which may have been written by people who had an interest in discrediting them, or they may have been dealing with mongrels they believed to be pure native.

Early on in your beekeeping you should briefly understand mating behaviour. In very simple terms the queen mates randomly with a number of drones, perhaps 10–15 normally. These drones may be any of those that are flying from the colonies in your area. I won't quote figures of how far drones fly because those I have seen vary considerably (with one source suggesting they can be found 20+ miles from the hive from which they emerged), but let's guess that the drones flying in any one area could possibly come from at least 100 colonies and probably many more. The chances of a pure mating are extremely slim unless you live in an isolated area where just one type of bee is kept and realistically this would be our native bee, or the area was flooded with drones of a particular type as the result of a serious queen breeding programme. In effect most of our bees are mongrels and a pretty mixed bag.

Different Types of Bee
As I see it, the main types available to us are as follows and the descriptions are based on my own experience and observations:

Carniolan (Apis mellifera carnica)
These winter as small colonies and rapidly build up in the spring, needing a large brood box because a single brood box National is far too small. I agree totally with their reputation for being very swarmy and this is backed up by speaking to many beekeepers in the countries where they are the bee of choice. I have known of many beginners who have bought a 5-frame nucleus of Carniolans that have swarmed a few weeks later. I have known both the original colony and the swarm swarming again in the same season.

Italian (Apis mellifera ligustica)
These are probably the most widely used species worldwide, especially in countries like the USA and Australia where their

climate is much more like it is where Italians evolved than ours is. They have yellow bands on the abdomen and are more distinctive than any other race. The queens have yellow bands, some being very light. They are very prolific with large colonies throughout the year. It seems to me that when there is no nectar coming in they simply turn any food they have into unwanted brood. To avoid swarming you must have a large brood box, certainly a double brood box National. On more than one occasion I have seen three used, although this may have been showing off. In my experience they need far more feeding than any other bees and may be in danger of starving during spells of poor weather during the summer. The only time I have known of colonies dying through nosema (apis) they were Italian and this has been experienced by many other beekeepers I have spoken to.

Italians have a reputation for drifting and robbing other colonies, but this may be because they are seen more easily than other bees owing to their distinctive colour. They are generally softer than other bees and often don't fly at lower temperatures.

Native (Apis mellifera mellifera)

These are the native bees of northern Europe and it is thought they colonised the UK over the Channel land bridge when the ice retreated after the last Ice Age, then became trapped when we were cut off from the Continent by the rising seawater. In the past they were referred to as the "British Black Bee" and acquired an unjustified reputation for bad temper, although this could well have been due to mongrelisation with imported races.

They are non-prolific and can be kept in a single brood box National hive all year, as they don't waste food on producing unnecessary bees. As they have evolved to be successful in sometimes harsh climates they are well suited to our conditions and they are thought to be the most adaptable of the major races, so are suitable for the whole of the UK.

The most pure native bees tend to be in isolated areas where the imported races do poorly. Many think they are extinct but only because they read earlier writings.

Imported Hybrids and Crosses

These can vary from controlled crosses that are quite reliable to anything that is available. It really is a lottery and I have seen both ends of the scale, with some being very different within the same batch. What the purchaser should be aware of is that as they have already been crossed, although they may be reasonably well-tempered they could be first crosses and the next generation may be quite vicious.

Mongrels

The vast majority of our bees are mongrels and this will probably always be the case. I have found that by setting some sensible criteria it is possible to produce mongrel bees to your liking that are consistent and reliable. As we get many opportunities to improve our bees during the active season it would be a good idea to take them.

Mongrels are sometimes seen as being bad-tempered, but many believe the continual importation of pure races, even if they themselves are docile, is one cause of the problem, poor handling being another.

I think mongrels are an excellent choice for beginners as they will certainly learn a lot due to variability, but my own preference would be for those that show the characteristics of native bees. These are termed native or near native.

Imported Bees

At the time of writing I understand all national BKAs have a policy of non-importation and that is done for a reason. I am totally opposed to the importation of bees and queens whatever the situation and support these policies. The biggest danger is obviously disease and although those who support importation will always argue that disease could come in some other way than on bees or by smuggling, my view is we should reduce the risk.

Fairly early in my beekeeping life I came to the conclusion that the imported races were not always suited to our climate.

In general they are far too prolific and Italians especially need far more attention than any others. I have no commercial interest at all, but I have known those people who discredit UK-raised queens to be the ones who sell imports, so you know their motives. Putting it simply I do not advise using imported bees or queens.

In case you think it is easy to bring queens into the country when on a foreign trip DON'T. We probably have varroa as a result of imports and at the time of writing there are other known nasties lining up to get at us and, as new pathogens are being discovered on a regular basis, plenty of unknown nasties too.

Obtaining Bees

This needs a bit of thought and sound advice as I have seen many problems. Beginners are often advised to buy a 5-frame nucleus of gentle bees, but I will argue very strongly against that. Why gentle? How is anybody going to learn how to handle a colony of bees if their learning is on some that don't move? If it comes from a commercial source it can mean the nucleus has been put together with bees from several colonies and given a Carniolan or Italian queen, probably imported, both of which will be prolific. In a short time it may be bursting with bees and not a situation I would advocate for a beginner who hasn't had the benefit of some good quality teaching and guidance. What I think a raw beginner needs is a well balanced colony that will expand naturally, not one that could explode.

I have seen some excellent nuclei supplied commercially and there will always be good honest suppliers, but equally there have always been those willing to supply bees to beginners with little concern for what happens afterwards. These people never seem bothered about reputation, simply because there is a regular stream of beginners and if there is a shortage an over-enthusiastic beginner won't take any notice of a warning to be patient. If there is a problem it is so easy for the supplier to say that the purchasers didn't know what they were doing.

One offering for sale that will always stick in my memory

is a 5-frame nucleus I was asked to look at that had no sealed stores, two frames of foundation, hardly any adult bees in it and was badly infested with varroa. How would inexperienced beginners know what they should be getting? On that occasion I guessed there might be a serious problem, so I took two other beekeepers with me. It's a good job I did because the nucleus was so bad that nobody would have believed me without witnesses.

Over the years, especially when there is an upsurge in interest, I have seen bee suppliers spring up. They are often very inexperienced people who see a fortune in front of them. Very often they take orders and deposits, then are unable to deliver because they don't know what they are doing. To avoid possible disappointment I suggest you do some serious enquiries before placing orders.

My advice to everyone before leaping in with both feet is to make absolutely sure you want to take up beekeeping and I don't mean just initial enthusiasm. Get along to your local BKA, handle full colonies of proper bees several times, not just those soft things that teach you nothing apart from the depth of your pocket. Learn the basics, speak to the demonstrators and good beekeepers and get stung several times so you don't need the usual get-out clause of being "allergic". If you have done all that and you are still enthusiastic then go for it. You may have waited some time, perhaps a year, but you have had hands-on experience and are in a very good position to handle a full colony entirely on your own, so it really doesn't matter how you start.

You should have acquired a hive and put it together. Wait for frames and foundation until you need them and get your stand ready. The options open to you are:

Swarm
"Oh, no! He's suggesting that beginners should take in swarms! He's crackers! Think of all the disease and bad-tempered bees that will be spread about!" I can hear the howls of protest as I write! I can see a few faces as well! If you are faced with that negative attitude simply ignore it, make a mental note never to take any advice from them again and

read on. You could be well rewarded in bees and experience others are denied. Don't forget that a swarm could have come from a colony of the person advising you against it, or perhaps your "adviser" sells bees so they may have a reason for guiding you away from a swarm.

A swarm will teach you an awful lot in a short time and in my view is an excellent and cheap way to start, but the least controlled as you may have to move quickly when you have been told about it. Swarms in general are very safe if dealt with in the way I describe under Swarms (page 107) in Chapter 8, Colony Increase, although of course there are the odd ones where there are problems, exactly the same as buying bees on combs.

Nucleus

A nucleus is a small well balanced colony on 3–5 frames. Bearing in mind the warning I have given I think it would be much better to source it locally. You may be lucky enough to have someone who is recommended by your local BKA, as some beekeepers produce nuclei to help beginners start. Your own BKA may have a scheme for helping beginners acquire bees. For guidance there is the BBKA "Standard Nucleus Guidance Notes" leaflet which gives good sensible guidelines and it is reasonable to expect commercial suppliers to comply with these. If someone is helping you out as a favour at much lower cost you must expect to receive a nucleus that doesn't comply and this could be that the queen is older or the bees need feeding, Even so any faults should be pointed out to you and it would be reasonable to expect help in getting it into good shape. It should be viable and healthy.

Full Colony

Unless someone is giving up, moving or downsizing you are unlikely to be offered a full colony, although you may be able to buy one at auction. As the auctioneer at West Sussex BKA auction I always inspect every colony in public. I inspect every frame and comment on it so the buyers have a good appraisal. The colony is also inspected for foul brood by the

Bee Inspector within two weeks prior to the auction, even
though I check as well. My only protection is a veil (as shown
in Plate 5) so potential buyers can see what the temper of the
bees is like. As our sale is at the end of April or early May very
often full colonies will be bought and immediately split, or
split throughout the season where the buyer can have at least
four colonies going into winter from the one purchased.

General

Whatever the situation if you are acquiring bees on combs
I think you should involve your BKA, as they will have
local knowledge and will be able to tell you if there has been
a history of sharp practice or disease with the potential
vendor. They will probably offer to inspect bees for you, as
the last thing you need with your first colony is to have to
burn it because it is diseased. There are many people who
work tirelessly on behalf of beekeeping and most don't
expect anything, but some form of "thank you" is usually
appreciated.

If buying bees on combs I wouldn't buy them later than
August if the price is quite high. Varroa and the higher
winter losses of recent years mean you may have a dead
colony in the spring that will be worthless. Before that you
have a chance of making sure it is as healthy as possible. If
the bees were reasonably priced it may be worth a gamble,
but seek advice.

I think it would be a reasonable request to ask what treatments the bees have had in the past. There is a legal requirement to record such information so it should be available.

However you buy bees I strongly suggest you use locally
reared queens, not imported ones. Ask the supplier if they are
home reared and don't be fobbed off with reasons why you
can't have them.

There are some parts, especially islands, that currently have
no varroa or foul brood. The importation of diseased bees into
some of these areas could make beekeeping virtually impossible, especially the more remote areas with harsh climates.
Please check with local beekeepers before buying bees from
outside and source from within your own island, not another

that is said to be disease free as it may not be. One selfish move could ruin beekeeping forever for others.

Moving Bees

Colonies have to be moved occasionally and you may need to understand how to do it at an early stage. Flying bees orientate themselves to their hive by recognising various features such as trees and bushes called markers. In good flying weather the foragers will normally fly around 1½ miles, often more if the forage is good, and will return to their own hive with incredible accuracy, so if you moved it 6 feet away they will still fly around the spot where it was. There is a useful rule that in good foraging and flying weather if you move bees it should be less than 3 feet or more than 3 miles. Bees can usually remember their location for about 3 weeks, so if you need to move them less than 3 miles in good flying weather you will need to move them twice otherwise they will fly to their previous location. This will need to be done by making the first move more than 3 miles from both locations, then moving them to their final destination after 3 weeks. It might be better to wait until winter when you can move them directly. If you need to move them a short distance in the same garden you can move them 3 feet a day; they will reorientate themselves in a couple of hours and return to the new position. If you need the entrance facing a different direction then gradually turn the hive as you go; 45° at a time will do. There are other tricks an experienced beekeeper will use, but for a beginner it is probably best to stick to the above.

In winter you can wait until there is a period of 3 weeks non-flying weather when you can move them as far as you like. To move a colony a short distance of half an hour or less, all you need do is wait until they have stopped flying in the evening and close the entrance up with an entrance block or something soft like hay, grass or sponge and secure them with a hive strap or baler twine and a turnkey after removing the roof and closing off open holes in the crown board. If you are travelling more than half an hour you will need to replace the crown board with a screen of some sort. Spray this with water to keep them cool. I have seen several colonies that have cooked following a journey without ventilation. The temperature rises quickly, the wax melts and the combs collapse. It is not a sight I would like anyone to see. If you are moving a colony 3 feet you can do it when the bees are flying by simply smoking the entrance and moving the hive.

Pheromones

There is no need for the ordinary beekeeper to have a detailed knowledge of the many pheromones that are present in a colony, but it is handy to know of their importance. Pheromones are chemical stimuli that are responsible for much of the behaviour of the colony. We often humanise bees and tend to put ourselves in their position, but they are incapable of thinking in the way we do. Instead their thinking is done by pheromones. It is thought we only know a fraction of the pheromones in a colony, their composition and functions. In some ways I think that is good for bees and perhaps the way it should stay. I dread to think what would happen if we could synthesise them all.

Examples are:

Nasonov Gland

This is often seen as a light coloured band on the last abdominal segment (tergite) of the worker where a pheromone is released. It is exposed and the bee fans air over it with her wings as a means of calling other bees.

Queen Substance

This is one of the pheromones produced by the queen and is known to have many effects including telling the colony if she is present or not. It is obviously very powerful because if you take the queen away from a colony they will be aware of her absence in about 15–20 minutes. Queen substance is produced by the queen in the mandibular gland and is dispensed over her body. She has a changing circle of workers called a retinue that are constantly grooming and touching her. In doing so they take queen substance either by contact or ingestion, then make contact with other bees and either exchange food in what is known as the food transfer pool or by contact. In this way the queen substance is dispersed throughout the colony.

5

ESSENTIAL EQUIPMENT

The appliance catalogues are full of all sorts of equipment, much of which you don't need or can improvise. With the importation of much of our manufactured goods from the low wage economies there is a lot of equipment that is poorly made. Things like hive tools and smokers are in your hands all the time and I have seen some so sharp you can cut yourself on them. Check with a sensible person at your BKA before buying anything and you should be given sound advice, as there is more than one way of getting stung in beekeeping.

At Wisborough Green we provide all the equipment needed before bees are acquired and that includes protective equipment, hive tools and smokers. We want to make sure everyone is happy to start beekeeping on their own before buying equipment and to be satisfied they are in a position to handle a colony competently without help and are not likely to cause problems for the bees or other people. I think all well run and responsible BKAs will probably do the same.

The minimum equipment required is:

Protective Equipment

I have a strong and well known view that the standard of handling bees in general is quite low and probably lower than when I started beekeeping. I believe the reason is because modern beekeepers are encouraged to dress up so they are completely protected against stings, with beesuit, wellingtons and gloves. The reason usually given is that it gives the beginner confidence, but in my experience all that happens is they don't get stung themselves, don't realise how many angry bees there are in the air or are stinging other people, so they don't learn how to handle a colony properly. It is so easy to get into bad habits that are difficult to get rid of. While writing this I have just had a telephone conversation with an experienced beekeeper who lives over 100 miles away and without prompting he complained about exactly the same thing.

I fully understand that people who are new to beekeeping may not be sure about their reaction when being stung, but sooner or later you will find out anyway. A good BKA will have a supply of protective clothing and it would make sense to fully protect yourself if you have any doubts. It gives you an opportunity to use someone else's equipment to see what suits you before buying your own.

It is normal to swell up when stung and if in a sensitive and soft spot it could be a large and painful swelling, especially around the head. I suggest you always wear head protection and if you buy your own I think a tunic with attached veil is an excellent choice to start with. Make sure it has elasticated wrists and waist, large pocket in front that is useful for putting things in and a removable hood and veil so it is easy to wash. Don't buy blind as the quality varies a lot and not always with the price. Go to a supplier or a beekeeping event with a trade presence and try all the gear on.

Some clothing is very well made from stout material that will last years, yet other clothing is little more than butter muslin. Remember that you will be spending a lot of time in it during the summer, when you will get hot. You will need plenty of room inside as you will be stretching and will probably encounter bushes with thorns so it should be tough. If it hasn't got a chin strap, then put one in yourself. I have seen many people with a frame of bees or full supers in their hand have the hood blow over their faces on a windy day.

The sizes stated are often one size smaller, so if you think you are medium then buy a large. I suggest you buy two, in case you tear a veil or want to let someone else have a look at your bees. For children we use small adults' full suits, as they will usually fit an 8-year-old and some 12-year-olds won't get inside a children's size, so there's a tip for parents.

You will find that trousers tucked in socks are far more comfortable than wellingtons.

Gloves

For these to be fully sting proof they have to be thick. If they are thick they are difficult to use, resulting in clumsy handling which can fire bees up. Any thinner gloves such as household gloves or disposable ones are far from sting proof and probably only have the benefit of keeping your hands propolis free. In my opinion the quicker you can dispense with gloves the better.

I suggest rings are always removed if possible as the majority of stings will be on the hands, whether you wear gloves or not.

General

I do not give the above advice lightly. I don't like getting stung myself and I doubt if anyone else does either, but you must accept that if you keep bees you will get stung and you must assume that every time a colony is inspected there is a possibility. There is nothing like a sting, especially if it was your fault, for concentrating the mind.

I am frequently asked how often I get stung and the answer is difficult to give. If I had one sting per colony that wasn't accidental I would be very disappointed. Very often I can handle 20+ colonies one after the other and not have one sting. That's when beekeeping really is fun.

The best ways to minimise being stung are to keep good-tempered bees and develop good handling techniques, not to get fully dressed up thinking you won't get stung, otherwise you may not develop handling techniques that will reduce it and you run the risk of others being stung.

At Wisborough Green at least 50 per cent of newcomers don't wear gloves and it's noticeable that they are the ones who usually go on to make good positive beekeepers.

Quite frankly if after a few hive inspections you are frightened of bees and being stung I think you should give up, as the fear probably won't disappear. It will be kinder on the bees and those around you.

Smoker

Smoking bees calms them, making the colony easier to control. There are those who advocate handling bees without smoke as all bees that are worth anything will object, I think this practice is highly dangerous and strongly advise against it. It is interesting to note that all those I know who don't use smoke always wear full protection.

There are now only the bent-nosed smokers available, which I think is a great pity, as I used one of the old-fashioned straight-nosed smokers for over 40 years and much preferred it.

Before buying a smoker I suggest you look at several. One major problem I find is the strength of the spring inside the bellows, as most I have seen are very strong and will make your hands ache. Some of the cheaper ones are of very poor quality and I have seen some with metalwork you can easily cut yourself on. Don't be fooled into thinking that as you only have a couple of colonies you only need a small smoker, as you may find a small firebox severely limits the fuel you can use.

Almost anything that will smoulder with a cool smoke and will stay alight will do for fuel including hay, dried grass and leaves, egg boxes, wood shavings and my favourite touchwood. Touchwood is wood that has been decayed by the action of fungus and is very light and easily broken up. If you have your bees in woodland then you usually have a ready supply of fuel, otherwise keep an eye open and have a sackful handy. If you collect wet fuel you can dry it in a greenhouse.

For some reason a lot of beekeepers have trouble lighting a smoker and keeping it alight, often because they fill it, then try to light the top! I use a piece of newspaper and light it before putting it in the firebox, then add something that will burn well such as shredded paper, dried grass or wood shavings on top. Keep pumping the bellows until there is plenty of smoke, add touchwood and anything else that will readily burn. This will give the fire some substance and all that is needed is filling up when it runs low. It is a good idea to put green grass on top of your fuel as it will prevent sparks being blown onto the bees, stop any loose fuel moving about and cools the smoke.

Don't lay your smoker on its side or it will go out quickly, keep it upright. When you have finished with it simply plug the hole in the chimney with grass or a cork and lay it on its side. Don't empty the fire box, but use the partly burnt contents to make lighting easier next time. Don't forget that a smoker can be a fire hazard so be careful before putting it in the car or shed and make sure it is cold. You also need to be careful in dry weather as it is so easy to set dried grass alight.

When inspecting a colony I always keep my smoker between my knees, gripping the bellows. You soon learn that a smoker with a wire cage around the firebox will prevent your knees getting burnt. Bees are good at telling you when you need your smoker. When they do, you need it in a hurry and it's easily available if it's between your knees, whereas if you put it on the ground you spend valuable seconds finding it, if you haven't knocked it over or it's gone out. You are also likely to get your face closer to the bees which is not a good idea.

The correct use of the smoker is one of the most important parts of beekeeping.

Hive Tool
For frames with long lugs I think the scraper type hive tool with a bent up end and a nail puller hole is the best, as it can be used for so many things. As with smokers, there are some poor quality ones available.

The flat end should be very thin as you will need to force it between the boxes in order to prise them apart. Some have a fat blunt end which makes them difficult to use and often damages the tops and bottoms of the supers because you can't easily slide them between the boxes.

There is another type with a curled end called the "J" type, which is probably better if you have short-lugged frames. See Fig. 6.

Keep your hive tool in your hand all the time as you will find it is constantly needed, so make sure it's comfortable.

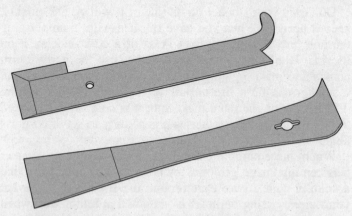

Fig. 6 Hive tools: "J" type hive tool (upper); scraper type (lower)

6

HANDLING BEES

All colonies need to be handled approximately every week during the summer if you don't clip your queens (see Chapter 7, Swarming, page 97), or every other week if you do, in total around 25 times a year for the former per colony, 15 for the latter, if you inspect as often as you should. That actually isn't very much, although if you have two colonies you can almost double that. Perhaps you will visit your BKA 10 times a year and you may handle a colony on your own half of those if you are lucky, so the amount of tuition is quite small. Unless you work with another beekeeper you will have little chance of learning from others, so your technique will be developed by yourself. Habits, both good and bad, are formed very early and the quicker that good handling techniques are learnt the better, before bad ones take hold.

Don't get near your bees unless you have face protection at all times. Many a beekeeper has ended up with unwanted stings when they didn't protect themselves properly because they "were only just going to . . ." Make sure your smoker is well lit and you have more than enough fuel for your inspection, bearing in mind that often the bees need more attention than you thought.

Bees are living creatures that I think deserve to be treated with respect. They have no legal protection like other animals have, but in my opinion some beekeepers don't treat them very kindly at all. In your manipulations make firm but gentle movements and if anything goes wrong, whether it is your fault or not, just close up as quickly and quietly as you can. I have rarely had an unexplained problem with bees where normally docile bees have been very bad-tempered and I think on each occasion I have the air pressure rapidly dropped. Whatever you do, don't panic.

Inspecting a Colony

Your colony records (see Chapter 13, page 143) should tell you if you need any equipment and this should avoid you having to leave an open colony to get something. When manipulating a colony you can work more efficiently if you have everything you need with you and you are well organised, especially if space is short.

Plate 6 shows a WBC hive during colony inspection. It shows the double-walled construction and the larger number of parts than the single-walled National hive has. The queen excluder is a wire one with a wooden frame. The right-hand hive of the two in the background is a National which is coated with paint that is able to breathe, preventing the hive from becoming damp.

The record sheet is on a clipboard and kept inside the roof so it is accessible. The supers on the left have not had the crown board removed, the one on the right which is actually a brood box being used as a super is temporarily covered by the roof, leaving only one face of bees exposed. Placing supers on the lifts in this way ensures few bees are crushed. There is a spare super on end that will be put underneath the others when the hive is reassembled.

This is a single brood box WBC hive with only 10 brood frames and is about to have the fourth super put on, where one of the others is a larger brood box. The super capacity of this hive is about 140 lb of honey. This shows that you can use hives with single brood boxes if you have non-prolific bees and they do produce good crops of honey.

It looks as if the floor is stood directly on the ground and the legs may rot quickly. It would be better to stand it on a concrete paving slab which would also allow free air to keep the hive drier, especially in the winter.

When inspecting a colony, approach the hive from the side so you don't disturb the flight path and always observe the entrance before smoking so you understand what is normal, or if there is a problem. If bees are busily coming and going and there is pollen coming in, everything should be fine. Bees returning to their own hive come in purposefully and don't fly directly at the opening, but land on the periphery first, then run in. If bees are flying around without any apparent purpose then dart for the centre of the entrance, it is possible they are robbers. An alighting board can often tell you something. If it is clear you generally have nothing to worry about, but a colony with heavy chalk brood (see page 136) will have the infected pupae on show.

Smoke the entrance with a few puffs of smoke. Unless the colony is known to be bad-tempered there is no need to wait "for the smoke to work". Stand at the side of the hive and put your smoker between your knees. Take the roof off, turn it over and put it on the ground at the back of the hive at 45°, so you can put your supers on it without crushing bees. Even if there are no supers it should be a habit anyway.

If there are no supers on the hive then put your hive tool in the gap between the crown board and brood box at either front or back of the hive where the frame lugs are. Push it in a reasonable amount without wiggling it about, as that damages the edges of the wood. Lift the end of the hive tool upwards so the point is pressing down on the frames. When you have enough gap, blow a few puffs of smoke into it. Hold the rim of the crown board and lever down on any frames that have got stuck to the crown board and are lifting up with it. Give smoke when needed. When the crown board is free, take it off and turn it upside down over the brood box to see if the queen is on it. Lay the crown board upside down against the front of the hive so it isn't blocking the entrance.

If there are supers on the hive there is no need to remove the crown board. If there is one super then do exactly the same as above, but put the hive tool between the super and queen excluder. Don't forget to use the smoker and put the super on the upturned roof. Remove the queen excluder by peeling it off from corner to opposite corner if it is a flat metal or plastic one, or in the same way as the crown board was removed if it is a wooden rimmed wire one. You wouldn't normally expect a self-respecting queen to stick around with smoke, but it might be wise to turn it upside down and give it a glance before laying against the front of the hive.

If there is more than one super you can please yourself if you take them off one at a time or all together, but don't leave the faces of bees exposed, cover them up with a spare crown board or roof. If you have several faces of bees exposed they seem to communicate with each other and if one group gets fired up they all do.

You may hear about cover cloths that are used to cover up the frames in the brood box that are not being inspected. I have never used and don't advocate them as I think they retard learning and can't see any real use for them anyway. You can learn so much by observing the behaviour of bees and you won't if you cover it up so you can't see it.

I normally start work at the near side of the brood box and work across to the opposite side. Occasionally I will move round to the other side to avoid leaning across the whole brood box, which can be tiring. I find it easier to see eggs and larvae in the bottom of the cells if I have good light, usually the sun, shining over my shoulder, so I will start from the most convenient side.

Before taking the first frame out of the brood box you will need to make room to do so, otherwise you may "roll" the bees due to the thickness of the combs varying. I use castellated spacers and I lift up the second frame in from one side so it just comes out of the spacer, then pull it gently towards me as far as it will go and rest it back on the top of the spacer. You won't damage any bees because the widest part is the thorax that is very tough, being basically solid muscle that powers four wings and six legs.

Do exactly the same with the third frame in and you should have created enough room to remove the fourth frame. By habit I generally look at this frame more than any other, as it should have brood in all stages, pollen and nectar/honey in it. If I see the queen on it I put it back in the colony and look at the next frame. If I don't see the queen I lightly shake it over the brood box to remove some of the bees, so I can check for foul brood disease (see Foul Brood, page 134.)

To shake bees off a frame I first look to see if the queen is on it, if so I put her in the brood box, if not I hold the frame by both lugs above the brood box, but diagonally so there is a greater length for the bees to fall on. Give the frame one fairly firm shake. I then lean the frame against the front of the hive and inspect the rest of the colony. When a frame is taken out of the castellated spacers I habitually move it away from the adjacent frame to avoid "rolling" bees between the combs when withdrawing it. When put back they can be rested temporarily on the top of the spacer, thereby maintaining a gap so I can remove further frames.

When inspecting a colony I do things by habit, and experience tells me when things are wrong, as very often they jump out at you. The colony is telling you something all the time, even if all is OK; and the "reading" of a colony is a very important part of beekeeping.

Plate 7 shows me inspecting a colony in a National hive at the Wisborough Green teaching apiary. Note:

- The supers are on the upturned roof. This gives point contact so few bees get crushed. They are close to the hive yet not in the way of manipulations.
- The area around the hive is uncluttered. This is a teaching apiary so there is reasonable space between the hives.
- The queen excluder and first frame to be removed are in front of the hive.
- There is little vegetation around the hive which keeps it dry.
- The hive stand is on a concrete paving slab keeping it level.
- The brood box and supers are "seconds" as can be seen by the small number of knots.
- The hive parts are branded for security.
- Holes in the crown board are covered up to avoid comb being built under the roof.
- The smoker is between my knees in case it is needed quickly.
- The hive tool is in my hand.
- I'm only wearing a veil for protection.

- The frame is being inspected over the colony in case the queen is on the frame and could drop off.
- My back is upright to avoid tiredness and aches.
- Despite it being a full colony there are few bees in the air showing that the bees are good tempered and handled well.

Some of the things I look for or which can easily be spotted when inspecting a colony are:

- Foul brood and any other disease.

- That the queen is present, but the bees usually tell me this as they give a roar and appear a bit agitated if they are queenless.

- If the queen is present is she laying? If there is plenty of food a queen would normally be laying during the summer, but when there is no nectar coming in or at the end of the season you can expect queens to reduce laying or to stop altogether.

- Is there brood in all stages? With experience you will be able to tell the ages of larvae. If they are hardly visible, but in a little pool of glistening white brood food then they have just hatched and are probably about 4 days old, counting from the day the egg was laid; if they are curled up and filling the cell and close to sealing, they are probably 8 days old. If you have both of these and eggs then provided that they are healthy, you have no problems.

   (Plate 8 shows a good brood frame with honey stored in the top corners as you get with non-prolific native type bees. The brood is largely sealed worker with a small amount of drone brood with the domed cappings at the bottom of the frame. It is sealed brood like this where you would find AFB and in the unsealed stage where the larvae fill the cell just prior to capping where you are likely to see EFB, so get into the habit of checking one frame with brood in all stages in every colony at every inspection.

   A frame taken from a brood box will usually be covered

with bees. This frame has been shaken to show the brood. The handler is not wearing gloves and has a hive tool in his hand, both good practice in my view.)

- Are there any queen cells? You should be aware that bees build queen cells under three impulses:

1. *Emergency*, where bees will convert worker larvae into a queen by extending a few existing cells to form queen cells. This will happen if the queen is no longer there, which usually happens due to something the beekeeper has done, e.g. killed the queen during manipulation.
2. *Supersedure*, that naturally happens towards the end of the season when the colony produces a young queen to replace her mother. In this case there is usually one, often two and occasionally three queen cells built.
3. *Swarming*, when five or more, but usually fewer than 20 queen cells are built. On one occasion I removed 119 from a colony!

Swarm and supersedure cells are built vertically and can be found anywhere on the combs, despite what we are often told. They are there for a reason and will usually have to be dealt with by the beekeeper, otherwise you could have one or more swarms emerge.

- Is there enough food to last until the next visit? This comes with experience and colonies do vary (see Chapter 4, Bees, page 67). I always like to see a reasonable amount of food in the brood box and if there is a reasonable amount plus some in the supers they should be OK for a couple of weeks. Good nutrition is important to a colony, not only honey but pollen as well. Food is needed for both adult bees and brood and this is not always realised by beekeepers. Poorly fed brood is much more susceptible to disease than that which is well fed. When food gets short the queen is fed less so she slows down egg laying or stops altogether and if a colony is desperate it will remove brood. If there is a problem in the summer you will need to look at all your colonies because if the shortage is weather related they may all have a problem, in which case you may need to feed syrup, otherwise you may be able to take frames of food from the more prosperous ones and give to those that are short, after shaking the bees off. If some colonies are short of food with no reason when others still have plenty, mark the queens down for culling.

- Is there enough room? This also comes with experience and a beginner will be surprised at the progress a strong well balanced colony can make in a week of good weather, especially if there is OSR or borage being grown in your area. It is quite common for a super to be filled within a week and what a lot of beekeepers don't understand is that nectar has a high water content when it is initially placed in the cells, meaning they need two or three times the amount of room for initial storage.

- Assess each colony. How does it compare with how it was at the last visit? How does its development compare with that of others in the same apiary? For temper and quietness on the comb see Colony Assessment, page 94.

When handling frames always do it over the colony until you know where the queen is, so that if she drops off the frame she will go back into the colony. Stand up straight so you don't get backache, because beekeeping is a hobby and we want it to be enjoyable. The more comfortable you are, the more colonies you will be happy to have – the more colonies, the more fun you have!

Unless there is a good reason for not doing it put the frames back in the same order. When you get to the last three, reverse the operation you did when making room to remove the first frame. Smoke over the tops of the frames to move the bees and replace the crown board or queen excluder and supers. If you move quickly you can put every-thing back by simply placing it down. In my opinion there is no need to put it on at an angle then twist, as you may kill far more bees that way, you just don't see them. You will always kill a few bees during an inspection (that is unavoidable) but keep it to a minimum.

Finding Queens and Seeing Eggs

Many beekeepers claim they can't find queens or see eggs and this is usually either a mental or physical problem. In order to look after a colony responsibly both tasks must be done on a regular basis and any problem must be addressed. The mental problem is often beekeepers' convincing themselves they can't find queens or see eggs, rather than adopting a positive attitude.

Fertile queens behave like the rest of the colony. If the workers are quiet on the combs, then so are the queens, but if they are very active and rush around, the queens will as well and these are called runners.

It is not easy telling someone how to look for a queen because we all have our own ways, but it is helpful to know what she looks like first. Fertile queens will avoid the light if they possibly can and this can help you find them more easily. When inspecting a frame there will be a gap where that frame came from that is exposed to the light and this I call the "light" side of the comb. If the queen is on the next frame she will quickly go round to the other side which is called the "dark" side, so when removing a frame have a glance on the exposed face of the next comb and if you don't see the queen look on the dark side of the comb you have just removed. I tend to look along the bottom of the comb first, then around the edges of the frame and finally across the frame either in decreasing circles or in lines. This is a sweeping glance, not concentrated and probably takes around 5 seconds. Then turn the frame over and do the same again; if unsuccessful go back to the first side. If you don't find her then, do the same with the next frame and so on.

Eggs are easier because they don't move so fast, but they are smaller. You will need to be in good light, if possible with it shining into the bottom of the cells from over your shoulder. It might pay to shake some of the bees off the frame, but look to see if the queen is there first.

If you have an eyesight problem I suggest you get it sorted

out otherwise you will always struggle. I used to have excellent eyesight, but since needing glasses I struggle to see eggs at the normal distance both with and without them. I bought several pairs of non-prescription glasses that I use specifically for seeing eggs. I keep a pair in several places and that has solved my problem. If that doesn't suit you, try a magnifying glass or Fresnel lens, both of which can be bought at a market stall very cheaply. If they don't deal with the problem little else will. Our Chairman has a problem with one eye and has a lens in his tunic pocket all the time. He knows he has a problem and has addressed it.

Colony Assessment
A colony will naturally change its queen when it swarms or supersedes its queen. We are able to change queens ourselves either by using queen cells that are raised by the bees, or by a simple activity referred to as queen rearing. Using either of these methods we have the opportunity to improve our bees but sadly many beekeepers don't take up that opportunity, often resulting in poor bees. Queen rearing is beyond the scope of this book, but all good beekeepers will look to improve their bees and there is no harm in a bit of planning at an early stage.

All beekeepers have colonies that are better than others, so a simple set of criteria that can be used in their assessment would be a good starting point, but you may need to handle other people's colonies in order to judge your own.

If you have learnt how to handle bees reasonably well you should be able to tell when they need more smoke and how much they need. A good-tempered colony will just need the odd puff, but one that is a bit stroppy will need quite a bit. Most colonies will panic and run a bit if they have great blasts of smoke directed at them, but they shouldn't run when smoked gently. Make notes of these characteristics as this is part of assessing your colonies when you inspect them.

We must assume that in any one inspection all the colonies have the same treatment, i.e. the same handler and under the same weather conditions, so if any behave differently it will allow you to build up an assessment of each colony, so that you will know which colonies to breed from and which to cull when the bees present you with queen cells. Both bad temper and running make inspections much longer and more difficult.

I have set out some simple criteria below that you can add to with experience.

Temper

I think this should be at the top of everyone's list. We mostly live quite close to others and it is bad news having bad-tempered bees. Temper is usually determined by the bees themselves, the handling of them or, as is quite common, both. Occasionally outside influences such as weather can cause bad temper, but for short periods only, in which case it is advisable to close a colony and inspect it later.

Bad temper can take several forms including unprovoked stinging, "meeters and greeters" that are waiting for you to arrive and "followers" that say goodbye and are still flying aggressively around you some distance from the hive. All of these are the responsibility of the beekeeper and are easily overcome. If your bees are consistently bad-tempered it's your fault, no argument about it.

Calmness on the Comb

I have already described "runners" that rush about all over the combs, making them difficult to handle and finding the queen. On several occasions I have found a queen outside the hive, although these are extreme cases. Running is not generally connected to bad temper, but it can be. In describing bees as calm, I'm not referring to those soft soppy bees that never move, but the ones who have some purpose and move gently in an unfussed way.

Prolificacy

False logic says that a strong colony with a prolific queen who produces great slabs of brood will get you more honey, but that is rarely the case. These prolific bees probably originated in areas where the weather is consistently warm and they didn't have to deal with the variable climate we have. In a warm summer there is no doubt they will often outperform less prolific bees, but in the much more common ordinary years when flying is difficult there are many more bees and brood to feed. They can use up what they have collected at an alarming rate and may leave the beekeeper with nothing. Even in a poor year I expect my bees to produce in excess of 50 lb honey per colony.

In a single brood box during the summer I would normally expect my bees to have an arch of food above the brood, or certainly in the corners of brood combs and possibly the outside faces of the outer brood combs to have little or no brood on them. If any of the brood combs have wood to wood brood I would consider the bees are too prolific for me.

If you take into account the much larger amount of sugar and the more management that prolific bees require, I think non-prolific bees are more suitable for the ordinary beekeeper.

Plate 1: A small group in a scheduled meeting at the Wisborough Green teaching apiary.

Plate 2: A flat-pack National hive for self-assembly.

Plate 3: Bees and children can mix.

Plate 4: A good site for hives.

Plate 5: A nucleus being inspected prior to auction (*courtesy of Neville Childs*).

Plate 6: A WBC hive during colony inspection.

Plate 7: The author inspecting a colony in a National hive at the Wisborough Green teaching apiary (*courtesy of John Glover*).

Plate 8: A good brood frame with honey stored in the top corners as you get with non-prolific native type bees.

7

SWARMING

Oh dear! For some reason this is a subject a lot (and I mean a lot!) of beekeepers, even some who have kept bees for years, just don't understand, or in some cases don't seem to want to. In fact it is very simple and once the process has been started by the bees is quite predictable. It is important to learn a few simple things and remember them. There are some uncommon reasons why a colony of bees will swarm, often unexpectedly, but there is no need to deal with them here. I will concentrate on the main reason which is because it is the bees' only means of reproduction. Every beekeeper will need to deal with it, so learning the process is vital.

In case you have a mental barrier, just look at it as simple mathematics with a little understanding of what the bees are trying to do. It's that easy! Swarming revolves round the life cycle of the queen and I hope you have already learnt that, as it's one of the basics you need to know along with the swarming procedure itself.

Many a beginner has had to deal with a newly acquired colony throwing off a swarm soon after its arrival, so be warned. Even a nucleus that's kept in a nucleus box too long will swarm if it's crowded.

During the summer all colonies build what are called play cups, which are the top end of a queen cell. This is normal and some of them should be checked by the beekeeper on every inspection. All the time they are empty there is no need to worry, but as soon as there are eggs in any of them the colony intends to do something and that is usually to swarm. You need to be in control from now on, otherwise you could have a neighbour with a chimney full of bees that you will have to deal with. I can tell you now that bees in chimneys are very difficult to remove, especially if they have been there a day or so.

Although there may on occasions be some variation the following is what normally happens when a colony decides to swarm:

- *Day 0 (Saturday)* The queen lays the first egg in a queen cell. There will be more eggs laid in other queen cells over several days, giving a succession. The first, however, will be the most advanced at all stages and in theory should be the first to emerge. This will be used to calculate everything from now on.

- *Day 3 (Tuesday)* The first egg hatches into a larva. Experience will allow you to judge the age of larvae as this will let you guess when they are likely to be sealed.

- *Day 8 (Sunday)* The first queen cell is sealed. Provided that the weather is good the swarm, which is called a prime swarm, will issue (i.e. leave the hive) with the old queen and some of the workers that vary in age including foragers, with many being young bees that may never have flown before. In the meantime the queen cells are developing and brood is emerging to partly replace those bees that have already swarmed. Once the swarm has left the colony, within 15 minutes the bees are still flying strongly and it is difficult to tell from the outside that a swarm has issued. If the weather is poor the swarm can be delayed several days until there is a fine day. This is probably why many, even some beekeepers, think bees swarm as a result of the weather rather than as part of a process.

- *Day 15 (Sunday)* The first virgin queen should emerge and one of two things will happen:
1. She will kill her sisters in their cells (or perhaps the workers help), will get mated and take over the colony, or . . .
2. She will go off with another swarm called a cast, but with fewer workers than the prime swarm, leaving the next

queen to emerge to do the same, or take over the colony. This may happen several times.

Now what's difficult about that? Why not copy it out and pin it on the underside of your hive roof?

Some other sources name the day that the egg is laid in the first queen cell as Day 1 which I think is confusing. For that reason I have called it Day 0 and I have given each one a day name to help you even further, not that eggs are laid in the first queen cell on Saturdays.

Once the swarm has left, it is entirely on its own. For that reason the bees gorge themselves on enough honey to last several days, which allows them to build new comb and survive a few days if the weather is poor.

Swarm Control

The above information must be learnt if you wish to be successful at controlling swarming, because the bees know exactly what they are doing and if you lose a swarm they are very firmly in control. If you do lose a swarm it is almost certain you will lose some of your honey crop.

There are many methods of dealing with swarming, but they should all be based on the procedure as laid out above. I think there is a lot more to swarming than many beekeepers think and there are many factors involved, some of which we may never know. In some respects I believe we are little further advanced than we were 100 or more years ago even though we have the benefit of the discovery of queen substance (see Pheromones, page 76) well over 50 years ago. If you ask any modern beekeeper what causes swarming they will usually come up with shortage of room and the fact that the queen is old, exactly the same as in books 100+ years ago. For some reason they don't want to acknowledge the role that beekeepers, queen substance or genetics plays.

There is no doubt in my mind that some bees are much more swarmy than others. In my experience it's sometimes difficult to keep Carniolans in the hive, so that should be one clue how to reduce swarming. I accept that overcrowding can be a visual clue although the amount of queen substance each individual bee receives may be the trigger, so give extra supers in advance of requirements. If a queen is so prolific there is wood to wood brood I would change the queen rather than provide more space in the brood box, although the latter can be a short-term measure. If you wish to use prolific bees on a long-term basis then you will need to consider multiple brood boxes or larger hives. If you have addressed all these issues and your colony still persists in building queen cells you will need to take action.

Before embarking on any swarm control system I would inspect every frame in the colony, shake off the bees, remove every queen cell that has anything in it, clip the queen (see Clipping and Marking Queens, page 104) and add a super. Return in 7 days and if there are still queen cells with eggs and larvae in you will need to take action. There are many systems of trying to deal with swarming, the most common probably being one of the methods of artificial swarming, the principles of which will be similar but with slight variations. I have never made an artificial swarm as a swarm control measure, but have helped many others and have used a modified version myself for other purposes. As it's so popular and known to most people I will suggest it here. You will need spare equipment as indicated, but if you inspected the colony a week earlier you have had notice that you will need it. I have modified the system slightly from the normal version to make it easier for a beginner. I assume you are able to find your queen.

1. Move the existing colony (A) to a new site in the same apiary where it will stay permanently. Distance doesn't matter.

2. In its position, place a new hive (B) with floor and empty brood box.

3. Remove the supers, inspect hive (A) and find the queen.

4. If the queen is on a frame of brood then put that in the middle of (B). If there are any queen cells on this frame then remove them all. If there are a lot of bees on it and you can't see queen cells then smoke the bees away gently rather than shake. It is better if there is plenty of sealed brood on this frame as it will provide young bees to nurse the larvae that will result from the queen laying in the new combs; it will give the queen more space when they have emerged.

5. If there were no supers on (A) then add a brood frame of food to (B). This is likely to be one of the outside frames. Put the side of this frame that was nearer the centre of the brood nest next to the frame of brood. This is important especially if there is brood on one side.

6. Fill up (B) with frames of comb or foundation. This is a major problem with an artificial swarm, as (B) very often builds swarm cells again when foundation is used. Drawn comb is less likely to cause this, but a beginner may not have any. Do what you can with what you have; it will be a learning opportunity.

7. Put the queen excluder and supers from (A) onto (B), close up and put the roof on.

8. You now need to fully inspect every frame in the original colony (A) where you may find several different situations that need dealing with in different ways

   a. If you have unsealed queen cells in (A) but no sealed, remove the more advanced ones that are close to sealing, close up the frames and replace those removed with frames of foundation, but at the side so you don't split the brood.

   b. If you have both sealed and unsealed queen cells in (A) then remove all the sealed and more advanced unsealed ones.

   c. If you only have sealed queen cells in (A) then remove all bar the best one.

   As the colony is queenless they may build emergency

queen cells on existing worker larvae, especially if you remove any swarm cells. The colony may swarm with the queen that emerges from your selected queen cell, but they will only normally do this if there is another sealed queen cell. In 8a and 8b you will have enough time for this not to be a problem, but in 8c you don't know how old the queen cell is so you will need to inspect the colony every 2–3 days to remove all emergency cells until the selected queen cell emerges.

9. Close the hive (A) up.
10. Notice I haven't mentioned feeding as a matter of course and I don't normally advise it, though as a beginner you will need to keep an eye on the situation. If you have non-prolific bees there should be enough food in the original brood box and the supers to tide both colonies over. With a prolific colony you may have little or no food in colony (A) if the queen has filled all the combs up with brood. For a week or so (A) will have few flying bees and lots of hungry mouths and brood to feed. You may need to feed one or both colonies, but leave it a day or so for the bees to decide which hive is theirs, otherwise you may start robbing.

From now on you have two independent colonies that are as near as we can get to a natural swarming situation.

Six to seven days later, depending on the age of larvae in the unsealed queen cells when you made the artificial swarm:

1. Inspect (B) as a normal inspection. They may have built more queen cells which is common, especially if you used foundation and/or you have "swarmy" bees. Shake bees off all frames and cut out all active queen cells. This colony may settle down now, but continue with 7 day inspections, or 14 if you have a clipped queen.
2. Inspect (A) and remove all swarm cells bar one good one that will provide the queen to take over the colony. A good queen cell will be heavily dimpled and a lightish colour. If it is smooth and fairly dark it may be dud. Remove all the emergency queen cells if they have built

any. Probably the best thing to do is to look through the colony first to see which queen cell you wish to retain, then shake most of the bees off all the other frames to make sure you don't miss any queen cells. Under no circumstances leave two queen cells otherwise they are likely to swarm with the first queen to emerge.

Let's see what has happened. The colony (A) intended to swarm, we have moved it away from its existing position, so it loses all the flying bees to (B) which has the old queen and food to keep it going until more comes in. You have left queen cells in (A) that should emerge in around 8 days' time or less, but you have returned a day or two before that happens to cut out any that could result in another swarm. The resulting queen should get mated and start to lay. If all goes well you have prevented a swarm and made increase in a planned way.

If you inspect (B) 7 days after artificial swarming and they are still building swarm cells then remove the queen and kill her, remove all sealed and the more advanced queen cells. Return a further 7–8 days later and remove all swarm cells bar one and all emergency cells. The resulting queen should emerge, get mated and lay.

You may read elsewhere about the Heddon method of artificial swarming. This is a modification of what I have described that involves placing (A) alongside (B) then moving it later to add more flying bees to (B) in the hope it will produce more honey. I have resisted suggesting it as in your early days it introduces complications I think are unnecessary and I suspect you would feel happier doing something simpler with a few less jars of honey.

As a beginner I think it is more important to deal with the swarming situation than worry about the quality of the bees, but in the not too distant future I suggest you assess your colonies and with more experience you can use situations like this for improving your bees by using queen cells from your better colonies.

You may have heard about clipping a queen's wing to prevent swarming. It does but only for a few days, then they swarm, but with a virgin queen. A portion of the wings on one side of the queen is cut off so she can't fly, or not very far anyway. In this case the swarm issues and clusters nearby waiting for the queen to join them, but as she is unable to fly she will drop to the ground and crawl around for a while. The swarm soon realises she hasn't joined them and returns to the hive. Sometimes they will find her and cluster around her, but often she will be found in a small knot of bees about the size of a golf ball. There are several possibilities and they include the queen being able to return to the hive, cluster underneath the floor, go in an adjacent empty hive, or cluster on something like a tree trunk or fence post where they will usually stay.

You can see that if you have unclipped queens and you inspect your colony today and find no eggs in queen cups you know you can't have a swarm for at least 8 days and that will only be if the queen lays eggs in queen cups within a few hours of the inspection. However, if you have clipped queens you can leave it 14 days, because even if the colony does swarm you may have lost the queen, but not the bees. You have the choice of clipping queens and inspecting every 14 days, or leaving them and inspecting every 7 days.

Clipping and Marking Queens

I clip my queens because it gives me longer between inspections. I mark all my queens because I need to find them quickly, but don't use the international colour code where there are five colours so you can tell the age of a queen if you use them. I only use yellow because I think it stands out better than the other colours and I see little point in buying more paint than I need. I don't think there is any need for most beekeepers to mark queens anyway, as I think it better they learn to look for a queen rather than a coloured dot, but I always advise clipping.

There are aids available to help with clipping and marking queens, but I always recommend that beekeepers do it by hand as I think it helps you to handle them. For 20–30 seconds drones behave exactly the same as queens before they play dead, so practise on them first as it doesn't matter if you damage them.

Keep at it and do several before clipping a queen, as it's not difficult to cut off a leg. I have seen several people clip a drone well,
then go to pieces when trying to clip a queen. This is something
your BKA should be able to teach you.

8

COLONY INCREASE

I would prefer beginners to know and understand how to make
increase from their existing colonies rather than buying more
bees. You will learn a lot more, it may help you with your
management and it will be cheaper.

Swarms
As soon as it's known that you keep bees the news will travel
fast locally and you will have a chance of a swarm if there is
one, either from friends and neighbours or your BKA. If you
want to use this route then you will need a spare hive with
frames and foundation ready, as there is often urgency needed
with a swarm. Don't put the foundation in the frames until you
need them, otherwise it will become stale and the bees may
not build it out properly.

   As indicated previously many people frown upon the use of
swarms as they seem to think they are all diseased, presumably with foul brood. In nearly half a century of beekeeping I
have only known of one swarm that had foul brood and that
was when a beginner took a swarm from a known EFB
hotspot, hived it on comb and fed it. Yet I have known several
cases where colonies have changed hands with foul brood.
Two cases in particular were when one commercial beekeeper
sold 50 colonies that were riddled with it to another beekeeper
and another case when an apple grower bought 30 colonies to
pollinate his orchards and about half of them had it.

It makes sense to assume that a swarm is infected with foul
brood unless you know its history; and I have no problem with
that approach. A swarm will take food with it and if it came
from a colony that is infected the honey will be infected too.
If you hive the swarm on comb, the honey will be stored and
used to raise brood and that may be the cause of further infection. If you hive it on foundation and don't feed it, the bees
will use the honey they brought with them to make the wax
needed to draw the foundation into comb and in a couple of
days it will probably have been used up. The queen will take
a day or so to start laying, but the resulting larvae won't need
feeding for another 3–4 days, by which time all trace of
infected honey will probably have gone. If the weather is good
there will be no need to feed at all and everything should be
OK but if the weather is poor then keep an eye on them. A
swarm doesn't have brood to feed for several days, so it only
needs honey to feed the adults and make wax with. I very
rarely feed a swarm, as in my experience bees usually swarm
when there is nectar coming in and they don't need it anyway.

There are a couple of disadvantages in using a swarm, but
they can be easily overcome. There is a possibility the queen
is old, but I don't think that matters as the bees should supersede her. There is a possibility they may be bad-tempered or
poor bees. However, all of these are still relevant if you
bought bees on combs and you can requeen them when you
have a chance.

Whenever you hive a swarm, it makes sense to hive it away
from your other colonies. A check after 3–4 weeks should
show up any problems and if there are any then deal with them
as discussed under Foul Brood (see page 134). In my opinion
it is much better that a potential problem is handled in a
sensible and controlled way like this, than to refuse the chance
of a swarm and have it end up with a careless beekeeper, or in
someone's chimney where it could be a source of infection for
some time that nobody knows about.

You may need to take the swarm yourself if you are in a
hurry, but for the first couple you take I suggest you have help
and advice. If you have taken a call yourself make sure they
are actually honey bees. If you are looking for a swarm then

gather some kit together and keep it handy. This could include a container to put the swarm in such as a cardboard or wooden box, an old sheet or sack without holes, several lengths of baler twine, saw, secateurs, queen cage and an old comb if you have one. Obviously you will need a smoker, hive tool, fuel, matches and protective clothing. Spare protective equipment is always handy as I often find that bystanders want to get closer and on occasions I have found them very useful for holding things like ladders. If you have joined a BKA you should be insured, but it would be advisable to check the conditions before collecting a swarm.

As a beginner you must look at the taking of a swarm in a sensible and serious way. It is easy to look at it as a source of free bees and forget the risks, but unless it is in an easy place you must ask yourself if you are competent enough to take it. If there is ladder work, are you happy working at heights? Is there something dangerous to fall on? Don't forget that if you have never been stung before you will need to keep calm, especially if you are up a ladder, as the chances are you haven't got a parachute strapped to your back and the last thing you need is to end up inside a greenhouse without having gone through the door. Yes, I have seen a ladder put up above a greenhouse!

I can't tell you how to take a swarm because I don't know the situation you may have. I have taken many hundreds of swarms and I am fully aware that the reader is likely to be a complete beginner and may not be able to guess what the bees are likely to do. If they don't look easy to take I would seek help. Most experienced beekeepers will give you a hand and you may be well advised to accept it.

If you do have to take a swarm on your own, then try to get as many bees as you can in the container and turn it upside down on your cloth. Prop it up with a stone or piece of wood. If you have got the queen, the bees will soon start fanning and will call the flyers down to join them. When they have, wrap the cloth over the container and tie with string so the bees can't escape.

It is a common occurrence for a beginner to go on a swarm collecting list just to get a swarm and when they have done so to refuse to collect others. Can I ask you to be reasonable and take further swarms even if you give them away?

Hiving a Swarm

As already suggested, hive a swarm on foundation and don't feed it. If your apiary is large then put the hive some distance from other bees. You don't have to go mad, but 20–30 feet is OK and don't forget you can move it 3 feet a day to the position you wish to put it when it is shown to be clear of foul brood. If you have a small garden or apiary and you can't get the new hive some distance away then have the entrance facing a different direction. You can rotate the hive gradually later.

Set the hive up on the stand with brood box, frames and foundation. Put on the crown board and cover up the holes. Put a sloping board against the front of the hive so the high end is against the hive. If you haven't been able to find the queen then get a queen cage and open it. The bees should have clustered in the top of your container, so keeping it upright untie the string, smoke gently inside and dump the swarm on the bottom of the board. The bees should run uphill into the hive, so scan over them, especially at the leaders to see if you can find the queen. Most of the bees should go in the hive within about 20–30 minutes.

It is quite common for the swarm to abscond the following day, but they will only be lost if the queen can fly. To prevent this you can put a queen excluder under the brood box for a few days until she is able to lay some eggs, then you can remove it, but if it is a virgin queen she will need to go out to mate. In general, fertile queens lay within 24 hours, so if there are no eggs after 2–3 days you may have a virgin queen and will need to remove the queen excluder. If you see the queen running into the hive and you are confident she is fertile you can clip her, or if you aren't confident at doing this you can cage her for a couple of days until the bees have built some comb.

Bait Hive

A swarm will send out scout bees to search for a new home several days before the swarm issues. They will probably have scouts looking at several sites and it seems the decision is probably made before the swarm issues. There seem to be certain criteria they prefer for a home:

- Where bees have been before.
- Somewhere they can defend easily.
- A cavity that is big enough for them to build a large enough nest, but not too big.

In practice a brood box is ideal and this can be set up as a bait hive, preferably in the shade. I don't scrape off any wax or propolis from any part. Close the entrance up to about the area of a matchbox on edge and put just one old frame in it. This could be an old discarded frame and it doesn't matter if it is a brood or shallow.

Sooner or later you will see the odd bee flying in and out of the bait hive, then more may come. After a few days there will be a lot of bees and you may think that a swarm has arrived. Very soon it will either arrive or the activity will stop abruptly and this may be because the beekeeper whose bees they were has prevented the colony from swarming, it has swarmed and they have been collected, or the swarm has found another home more attractive than yours.

If you do get the swarm, then shake the bees off the old comb and fill up the brood box with frames of foundation and don't feed. Do this as quickly as possible otherwise the bees will store honey in the comb and build wild comb in the gap where there are no frames. It would make sense to clip the queen or put a queen excluder under the brood box until the queen is laying.

If you put a queen excluder under the brood box you should have a look at the colony after 4 days. If the queen was fertile she should have been laying and there is little chance of the swarm absconding; if there are no eggs the chances are you have a virgin queen and she will need to get out to mate. Either way you should remove it.

It may be difficult for beginners to set up a bait hive because they may not have an old hive or comb, but a request to another beekeeper may bring a positive response.

Splitting Your Colony

I think a practical person who has handled a colony several
times should be able to make increase provided that he or she
fully understands what is happening and doesn't simply
follow the instructions, otherwise that can result in failure.

There are several ways of making increase depending on
the state of your colony, but I wouldn't suggest that a beginner
attempts it unless it was filling the brood box. You should take
two things into account: firstly you need a queen or means of
making one; and secondly if you move bees in an apiary the
flying bees will go back to the original site.

I give some options below and in each of them you should
make sure both colonies will build up strong enough to get
through the winter, so I wouldn't do this after the middle of
July without guidance. Be prepared to take frames of brood
from one colony to give to the other, but shake the bees off
first. This may be done over a period of several weeks, but be
aware there may be initial imbalance that will show itself
later, for example, if flying bees are bled off one colony into
another, they are older and will die quicker than those left
behind. If I add frames of brood I like to use sealed brood
because it doesn't need as much warmth and attention as
unsealed brood and it will emerge earlier, which adds to the
adult population quicker and gives the queen room to lay in.

I give three simple options below, together with variations
depending on whether you have queen cells available or not:

Option 1

If your colony is building swarm cells you can make an artificial swarm as detailed in Swarm Control, page 100. If you use
foundation to fill up the brood box in (B) it will build up
slower than if you use comb.

Option 2

Exactly the same as Option 1, except that instead of only
putting one or two frames in (B) you put three or four, but
leave the supers on (A). As you haven't got the food in the
supers you will need at least one good frame of food; the ideal
will be one frame of largely sealed brood and one frame of

both sealed and unsealed brood. It doesn't matter where the queen is, or if she is known to be old you could leave one queen cell in each colony and deal with the queen cells accordingly.

Option 3

This is taking a nucleus off a full colony assuming there are no queen cells. You will need a queen or a means of making one, but for a beginner I would only consider a queen, as I think that raising a queen for yourself will give you a bit too much to think about, unless you have guidance. Ask other BKA members for a spare queen as during the summer there are usually plenty available. Tell them what you want to do, so they give you one from good stock, not just a makeshift to get you out of trouble. A queen can be introduced in the normal way (see Queen Introduction, page 146). Once you know there is a queen available you can start.

If you have somewhere more than 3 miles away to put your nucleus you won't have problems with flying bees going home and depleting it, otherwise you will have to add young bees that have never flown.

You can make a nucleus at any time of the day, but for a beginner who was going to keep it at home I would prefer the morning as it will give the flying bees time to decide where to go and the beekeeper an opportunity to strengthen a depleted colony. As with all small colonies, close the entrance down so they can defend it. I never feed nuclei as a matter of course, as my colonies usually have a good supply of food in the brood boxes and I can give one of these frames to a nucleus rather than syrup. Why make work for yourself when there is no need to?

Option 3a

This is if you can take it more than 3 miles away for 3 weeks. If you have a nucleus hive or can borrow one, then use it as it will be easy to carry, otherwise use a full hive. If you wish to make a 3-frame nucleus, then take one frame of stores and two frames of largely sealed brood, all well covered with bees and the queen. If you are not bothered about getting much honey it wouldn't harm to shake in another frame of bees. Fill the box up with frames of foundation. Make sure the frames are spaced properly otherwise you may kill the queen in transit. Take the nucleus to the temporary site and open the entrance.

Fill up the brood box of the full colony with frames and introduce the queen.

If a nucleus is made up like this and there is a honey flow it can build up very quickly, so be careful if you leave it in a nucleus box.

Option 3b
This is if you leave your nucleus in the same apiary. If you haven't got a nucleus hive, then set up a full hive on a stand in the position where you want it to stay. Make up your nucleus in exactly the same way as Option 3a, but leave the queen in the full hive and shake in the bees from two extra well covered frames. If this is done in the morning, you can have a look in the afternoon to see if there are still enough bees left covering the frames. As a guide, the frames should be as well covered as those in the full colony. Introduce the queen.

As there won't be many foraging bees for a week or so I would check on the food situation if the weather was poor. If it needs food I would prefer to give it a comb of stores from your main colony.

9

SUPERING AND DEALING WITH THE HARVEST

Supering

If you have acquired second-hand equipment from a known clean source you may have drawn comb. If so, make sure the frames are properly nailed together and clean them up, as I have seen many fall apart when being removed from the hive and in the honey extractor where they can't withstand the centrifugal force. If there is old pollen that is hard, then rake it out with the turned up end of your hive tool so it is cut back to the midrib. The bees should repair it if you can put these areas next to a comb that is fully built. If you have new equipment you will probably have foundation, but don't put it in the frames until you need to use it.

Brood combs are built out much better above the queen excluder than in the brood box, so I use a number of brood boxes fitted with foundation as supers. When the combs are built out and filled I extract them and use them in the brood boxes. I suggest the beginner does that at an early stage because you will find you will need brood combs if you wish to have a second colony. Conditions are not always right for building combs from foundation so a ready supply of brood combs will be very helpful.

I add supers when the colony needs them and this comes with experience. It is probably better to put on your first super too early rather than too late, otherwise you could have a swarming problem. When the bees are just occupying the outside frames is probably about right, but early in the spring a colony can make great strides in a week of fine weather, so be careful.

If you have a flat queen excluder, smoke the bees and scrape the wax and propolis off the tops of the frames and the brood box. Smoke the bees and lay the excluder on the brood box, I put the slots at right angles to the frames, but I very much doubt if it makes any difference. Then put on a super and crown board on top, making sure the holes are covered up, otherwise if the bees run short of space they will build comb in the gap in the roof.

Until harvest there is no need to remove the crown board at all. To add supers I always put them underneath the full supers. You can see the amount of space the bees have by simply tipping the supers up to have a look, as well as judging their weight. Until the middle of July I put supers on in advance, but, after that, swarming is unlikely to be much of a problem and the bees will soon be putting food in the brood box as the queen reduces egg laying, so less room is needed above the excluder.

Dealing with the Harvest

Even in your first year you should be prepared to have a crop, as that well known phenomenon beginner's luck extends to beekeeping as well, especially those who grasp the basics quickly and don't do anything silly. Many beekeepers find the removal of honey and its processing one of the most tedious parts of beekeeping. It's true that naturally messy people will get honey in places where you wouldn't normally expect to find it, but with a bit of common sense and planning it should be problem free. By using the word "processing" I don't mean in the way we have become accustomed to where food is over refined, but the procedure involved in getting it from hive to mouth.

There are generally two ways of producing honey: in the comb and as liquid honey to put in jars. At a later date you may wish to do both and some do. I will only briefly explain the procedures here, but you need to know which one you wish to do as soon as you get your bees, as there is a different approach to each. If you have acquired your bees from another beekeeper you may well have some shallow super frames; if so it would be better to work for extracted honey until you decide what you wish to do. Otherwise you can please yourself, the bees won't mind at all.

Extracted Honey

This is the most common method of producing honey by a long way. The combs have their wax cappings cut off, called uncapping, and the uncapped combs are put in a honey extractor which is a circular drum with a revolving cage inside, that is either powered by hand or an electric motor. There are two kinds of extractor: radial where the combs all face inwards like the spokes of a wheel and both sides of the frame are extracted at the same time; or tangential where the combs are placed against a wire screen facing outwards. They have to be turned over as only one side is extracted at a time. I wouldn't worry too much about the technicalities of honey extractors as many BKAs will rent or loan all extracting equipment to members and you should be shown how to use it. Get practising on someone else's extractor first, as there are many different kinds and many beekeepers have initially bought a brand new one, then quickly decided they didn't like it and bought another new one. What an expensive way of deciding on something you are only going to use a couple of days a year! There are several areas in beekeeping where it will pay to wait your time and the purchase of a honey extractor is one.

The combs are uncapped in some sort of tray and this can be heated, so the wax melts allowing it to separate from the honey, or cold where the honey needs to be drained from the wax. The latter is preferable if the honey is liquid, but granulated honey may need to be warmed.

The foundation used for combs that will be extracted is usually wired to avoid breakages due to the force of extracting, otherwise if you use unwired foundation you will need to be more careful.

Having got your honey extracted you still have several

other stages. You will need to strain it and decide how you are going to pack it. If you wish to sell it, even if only at the door or to friends you will need to comply with current regulations and these will be available elsewhere. If you only intend your honey to be used for the family or gifts, it should still be presented well.

When you extract honey you introduce air into it as well as particles of bees, propolis and wax. Although these won't harm you and may even do some good, some consumers may think it is "unclean", which is daft really as they will eat wax if they eat comb honey. It is possible to deal with these at the same time by using what is known as a bottling or settling tank, which is basically a round drum with a tap at the bottom. Some have a metal strainer on top which is usually referred to as a coarse strainer. If you want it strained finer, you can use a strainer cloth as sold by equipment suppliers, or a piece of plain nylon curtain material, which you can place over the top of the bottling tank. If there is no clamp to hold it, take a piece of strong string such as baler twine and tie it so you have a loop that is bigger than the bottling tank. Use a turnkey to tighten it, but before it's too tight push the centre of the strainer cloth down several inches to make a depression, then tighten and secure the turnkey. This makes a nice cheap strainer that can be washed easily and will last for a long time.

You will have to get the honey from the extractor to the bottling tank in some way and that isn't always easy. Honey will be more liquid if it is warmed gently, but some people don't like doing this as overheating will damage it, although there should be no problem if you are careful. Some honey will be granulated, or starting to granulate and won't go through the strainer and the only way you can liquefy that is to warm it. I find a preserving pan ideal. Tip the warmed honey onto the strainer and when you have finished leave the bottling tank in the warm for a couple of days. This will allow the air bubbles to rise to the top and you can take clear honey from the tap. Otherwise if you jar it immediately you will have air bubbles in every jar and although this won't harm you it doesn't look very nice.

Honey has natural yeasts in it and it may ferment if the

water content is high and it's kept in the warm. For long-keeping you should aim at below 18 per cent water and you can check it with the use of a refractometer. There are two ways of getting a high water content: if it is there already where the bees haven't driven off all the water (and this may be before sealing), or because honey is hygroscopic and exposure to air will quickly allow the honey to take up water (which could be because it has been left in an open container, or the lid of a jar has been left loose).

I'm assuming in your first year you will want to put all your honey straight into jars, though when you get a bigger crop you may find it easier putting it in plastic honey buckets with airtight lids, where you will find storage easier. If you put your honey straight into jars it may granulate quickly in which case the crystals should be fine; if it granulates slower, the crystals will be coarse. The rate of crystalisation depends on the balance of sugars, which is beyond our control and the temperature at which the honey is kept, which isn't.

To keep well, honey needs to be at a constant cool temperature. Around 10–12°C is suitable and at this temperature fermentation shouldn't be a problem. With central heating it could be a problem keeping it this cool, but if you have a cellar that should be ideal, otherwise a larder. The loft, garden shed or garage is usually a problem, especially in the summer as the temperature will fluctuate and fermentation will be a big problem, but only if the water content is high.

Honey jars are available from equipment suppliers, or perhaps from your BKA, but for small quantities you can probably make do with jam jars. They should be washed out and odourless, but don't use pickle jars.

If you need to liquefy granulated honey, just loosen the lid and warm it gently. In the right conditions a windowsill, conservatory, or greenhouse work well. If you don't liquefy all the crystals it will quickly granulate again.

I have no experience of ling heather honey but I certainly like the flavour of it. It is gelatinous and needs different treatment from what is termed "flower" honey. Colony management and extracting techniques are different and if you are in a heather district you would be well advised to seek guidance from a local beekeeper. OSR is usually an early crop and can granulate rapidly, especially if the weather is cold, therefore it needs extracting very soon after capping. Even if you think you haven't got OSR in your district I suggest you keep an eye on your colonies in May when it is in flower, as the bees will fly some distance to forage on this crop. Don't leave honey in the extractor or bottling tank longer than a few hours as it will granulate and won't pour.

Comb Honey
I think it's a pity so few beekeepers produce comb honey because when the flavour of the honey is good it rarely stays uneaten for long. Clearly when OSR is in flower it will be granulated, not liquid, but there are many who like it granulated. The good thing is you don't need any extracting equipment; the bad thing is that storage can be a bit of a problem.

There are two ways of producing comb honey:

- Sections. Traditional sections are small square wooden boxes that are fitted with wax foundation. They are put in a special super where the bees build the comb, fill with honey and seal them as in a normal comb. There are also round plastic alternatives, but sections are not much used because the bees need good nectar flows in order to make good sections and our climate is not reliable enough for that.

- Cut comb. This method has largely replaced sections, where you use unwired foundation in standard frames in standard supers. The comb when ready is cut up. Foundation used for cut comb is thinner than that used for extracted honey, so order the right kind.

You can either put full sheets in a frame, or what are known as "starters", which are strips of foundation about 20–25mm wide that bees will draw comb on and extend. Use narrow spacing for cut comb.

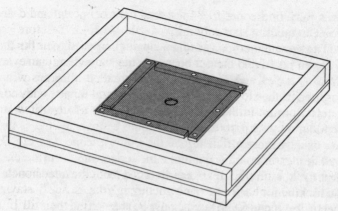

Fig. 7 Simple home-made clearer board

To harvest cut comb, all you need do is wait until the frames have been fully sealed, then remove them and shake the bees off. Lay the frame on a flat surface and, with a sharp knife, cut round inside the frame leaving a small witness of the comb and remove the frame. The comb can be cut up in whatever size you wish, but you may find the cut comb containers sold by appliance dealers convenient and clean, so cut to suit them. If you are going to consume it quickly, there should be no problem, otherwise store it in the freezer for a week. This will kill wax moth in all stages and avoid possible embarrassment, where what was probably an egg has turned into a wriggling larva. The frames can be returned to the colony for rebuilding and filling, hence leaving a small piece of comb in the frame for the bees to build the next comb on.

Clearing Bees From Supers

Bees need to be removed from supers so the frames can be extracted and there are two simple ways to do it.

The first is to take each comb out of a super, shake as many bees off as possible, brush off the remainder and put them in an empty super. I don't like bee brushes as I think they annoy bees, so I use a thin green twig from the hedgerow and take all the leaves off. You will need to be quick about this otherwise bees may find some free food that isn't being defended and you can quickly start robbing.

The second method is using a clearer board. Remember that a crown board and clearer board are the same thing, but when used as a clearer board it has what is called a Porter escape inserted in the slot. This is a plastic or metal device with non-return springs in. There are other designs of clearer board without moving parts and these should be investigated as they are usually more efficient than the Porter escape. See Fig. 7.

The clearer board is put under the supers and the principle is that the bees in the supers are divorced from the queen so they go looking for her, but having got through the escape they can't get back. Of course no bees should be able to find their way into the undefended supers otherwise you will have a serious bout of robbing and when this happens bees and wasps can clear a couple of supers of honey in a very short time.

When using clearer boards the rate that supers are cleared of bees is largely temperature-related; the warmer the weather, the quicker they work. However, in general if you get the clearer board on in the evening you should be able to take the supers off the following morning. Clearer boards don't work very well if the colony is queenless.

When you have cleared your supers, get them into where you are extracting your honey as quickly as you can. Don't leave any doors, windows or extractor fans open otherwise you could soon have some help. Don't let bees out of the window or they will go back home and tell their sisters where there is some free food and you could have a big problem. Kill them or tolerate them until you have finished as they usually stay on the window pane and out of your way.

It is much better to shake bees off frames if you are doing cut comb.

At the time supers are taken off the hives, some colonies may have little or no food in the brood box. Check and feed if necessary.

Cleaning Supers
Once the honey is extracted you will want to know what to do with the supers. If you need to put them back on the hive

for the bees to fill up, as you would after extracting OSR honey, then do it at dusk, otherwise you will cause excitement if you do it when the bees are still flying and this can cause robbing. If you have extracted the main crop, you will only be putting the combs back for the bees to clean up ready for storage. If you simply put them back on the hive, the bees will clean them up and probably store the honey in the middle of the lowest super. To avoid this you need to fool the bees into thinking they can't defend it, so I take a piece of thick plastic slightly larger than the brood box and cut a hole in it just big enough so you can get a finger through. An animal food bag is ideal. Put this on the brood box so the hole isn't over a frame and put the supers back on over that. A colony will clean up about 6 supers overnight, but you can leave it a couple of days before removing them if you have to.

Storing Supers

Good super combs are valuable items, but unfortunately many are neglected. Mice and rats are major problems. Rats are usually only a problem if the combs are stored wet with honey, or they are exposed in some way. Mice are probably the worst problem as they will quickly destroy combs to build a nest in.

Stack supers with a queen excluder top and bottom so mice can't enter. Keep them cold to avoid possible wax moth problems. They can be stored outside with a hive roof on the top, but not so high they will be blown over by wind. If stored under cover there is no need for a hive roof.

10

FEEDING

As far as the beginner is concerned there are two reasons for feeding: to avoid starvation in the spring and summer; and in autumn to replace what has been harvested. The food is usually ordinary white granulated sugar made up into syrup. Traditionally there have been two strengths of syrup, thin for spring and summer feeding and thick for autumn feeding, but apart from my early days I have only ever used thick syrup, as my bees rarely need feeding to avoid starvation and if they do I give them thick.

To make syrup the old measures were 2 lb of sugar to 1 pint of water, which in metric is 1 kg sugar to 625 ml water, although you can get it considerably more concentrated than this. I find an easy way to make it is to use a plastic bucket with a spout and these are often obtainable in markets very cheaply. Put 2 kg sugar in the bucket, add 3.75 litres of water and stir with a stick until it is nearly dissolved, then add another 2 kg sugar and stir until that has nearly dissolved, then finally another 2 kg sugar. Constant stirring is unnecessary, just occasionally will do until it is completely dissolved. If you put all the sugar in at once the extra effort needed to stir will make your arm ache. You can use hot or cold water but hot will dissolve the sugar much quicker. Ordinary white granulated sugar is suitable.

If the weather turns cool for a week or so after the OSR crop is taken off in the spring, the bees may need feeding to

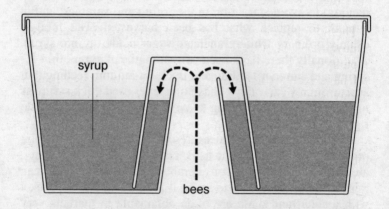

Fig.8
Plastic contact feeder (above)
How a rapid feeder works (below)

avoid starvation, especially the more prolific ones, where the queens lay wood to wood brood and there is no room for food. I find that the less prolific bees I keep are normally safe, because they usually have a reasonable amount of food in the brood box at all times.

There are several kinds of feeder and probably the most popular is what is known as a contact feeder. This is a plastic bucket with a tight-fitting lid that has a fine metal gauze moulded into it. The bucket is filled, the lid put on and turned upside down over a container to allow a small amount of syrup to run out, which creates a vacuum. It is then placed over the feed hole in the crown board and the bees suck the syrup through the gauze.

Another type of feeder allows the bees limited access to the syrup by climbing up a tube or channel. There is some sort of baffle that prevents the bees having full access to the syrup and drowning. These are called rapid feeders and come in several types, varying from small ones of about 4 pint capacity (a good choice for the smaller beekeeper) to larger ones that cover the brood box (more suitable for the larger beekeeper). See Fig.8.

The normal time to feed is when the main crop has been removed and the supers put back for cleaning in August or September.

I have seen written down many times how much food a colony needs to get through the winter, but you are rarely told it depends on the size of the colony or the type of bees kept, both of which are significant. In my experience a full colony of Italians will need more than double the amount of a full colony of native type bees. I feed mine until they won't take any more.

To prevent the bees getting excited and possibly causing robbing, the first feed should be done at dusk, allowing the bees to quieten down by the morning.

When the weather is cold, bees won't take syrup, so to avoid starvation you may need to give them frames of food from another colony and put them next to the cluster, or feed candy which is available from equipment suppliers.

11

WINTERING

Some bees are capable of wintering in quite harsh climates, but others are not. I am aiming this book at beekeepers in a very small part of the world that is less than 1,000 miles from north to south, yet weather conditions are totally different. In the Scottish islands such as Orkney they have periods of three months solid in winter when bees are unable to fly, yet in Cornwall they can have outdoor daffodils in January. All bees will probably survive in Cornwall, but not in Orkney where the native bee is the most successful.

Bees store food as far away from the entrance as they can, both above and behind the brood nest, presumably so they can defend it easily. Depending on the type of bee, the queens may reduce laying or cease altogether. Bees cluster together to keep warm, closer when colder, more open when it is warmer. They will gradually eat their way across and above the food in the combs after removing the cappings. When the weather is warmer, they will break cluster and forage within the hive, bringing food from the flanks into the area they have taken food from. In periods of warm weather some bees will make short flights to defecate.

If the queen has stopped laying for some time, she will start again, probably in January and the bees will have to raise the temperature of the brood nest which uses extra food. The resulting young bees will replace those that have died and the numbers will gradually rise. February and March are critical times for a colony as the older bees are dying off at a high rate and food is being consumed much faster than it was a few weeks earlier.

In order to winter well a colony needs the following:

Bees
Winter bees are reared in the late summer and autumn and need to be well fed as larvae and healthy, in order to bring the colony through the winter. Varroa and its vectored viruses can take great toll on them and shorten their lives, causing them to die during the winter or early spring, which weakens the adult population when they are needed for brood rearing, so you will need to monitor and treat if required.

Queen
She needs to be fit and healthy, if not she could fail and be unable to lay well in the spring. It is quite common for queens, even young ones, to become drone layers and if this happens the colony is usually doomed. Some queens are superseded by the colony in the autumn and very often you will find an old queen and her daughter on the same frame in the spring.

Food
This is both honey and syrup liquid stores to feed the adult bees and pollen for early brood rearing in the spring when there is little natural forage. If all the food is granulated, the colony is unable to use it, as the bees can't liquefy it as fast as they need it and the colony can starve with food on it. This is most likely when ivy honey, which is often collected in large quantities late in the season and granulates solid, is stored without much other liquid food among it.

The good beekeeper will make sure that the colony has enough food by feeding well in the autumn (see Chapter 10, Feeding, page 125). Hefting will give a good idea of the situation throughout the winter and spring. This means gently lifting the colony up off the stand by each side and the back several times during the winter, to compare with the results last time. You do it each side in case the colony is to one side of the brood box. This does need to take into the account the weight of the hive and the point of balance. Modern hives are usually about the same weight, but older ones can vary a lot depending on what wood they were made from.

As already mentioned, I have the feed hole running at 90° to the frames all year round, so in winter I can see how deep down the cluster is and if there is any sealed food above it. If it was parallel with the frames, the hole is immediately above the central frame and you can't see the comb. If the weather is cold and you can't see the cluster, you can put your ear to the feed hole and give the crown board a sharp tap with your knuckles to see if the bees buzz. Dead bees don't buzz and they don't collect honey either. If a colony dies from starvation, not only has a colony been lost, but also the food it went into the winter with. Starvation is usually the fault of the beekeeper.

Dry Hive
Dampness is a bigger killer of bees than cold. Check all roofs to make sure they are leak-proof. If there is a problem, they can be mended by warming propolis in your hand until it is putty consistency and plugging the hole with that, or using silicon sealant. In both cases the area should be dry. Through ventilation is always helpful. I use OMFs and remove the cover from the feed hole (only one if there are more) and this keeps the hives dry. I don't use matchsticks under the crown boards as suggested by some as my method has half the area exposed and that is adequate.

Vermin Free
About the only other problems are likely to be mice and green woodpeckers (yaffles), which aren't present in Ireland.

Mice will get in a hive when the bees have clustered and will build a nest where they have warmth and food. They will ruin combs and disturb the bees, which will probably consume more food than normal and may not survive. A shallow entrance or a metal mouse guard put in place in early November will prevent problems with mice.

Green woodpeckers can be a problem in woodland and will make a hole in the side of a hive in minutes; of course, they are after the bees inside. There are various ways of stopping them: using a cage made from wire netting which can be difficult to use and store, or thick plastic sheet such as DPC pinned to the outside of the hive with drawing pins, so the woodpeckers can't gain a foothold. They have good memories and can remember from one winter to the next, so once you have a problem you will need to deal with it in subsequent years.

12

PESTS, DISEASES
AND ROBBING

In the past beekeepers used to ignore any information on pests and diseases assuming they wouldn't be affected and any lectures on anything related to them were very poorly attended. The modern beekeeper is unable to do that because some diseases can rapidly result in a dead colony and recently any presentation on the subject has usually been well attended. Any responsible BKA will make sure that members, especially beginners, will have the latest information.

Pests and diseases are a complex issue and although there are many things we still don't know, it would pay to learn as much as we can about them and their treatments, as information is constantly changing due to research and legislation. There is a lot of good information available and it can quickly be accessed via the internet, but make sure it is relevant in your country, especially treatments. You should take current advice from official sources and I will only deal with them briefly here.

There is now so much awareness and concern about treatments, especially chemical ones and that is understandable, but the softer alternatives are often not so effective, resulting in the beekeeper not only needing to understand the disease, but what the treatment is trying to achieve.

Some diseases have been with bees for a long time and they have built up resistance to them, some sub-species more than

others, but some pests and diseases have recently jumped
species and there is little resistance to them by our bees,
although some beekeepers are reporting success with bee
breeding programmes on a small scale.

We can largely put them in two groups, those that affect brood
and those that affect adult bees, although some affect both. I will
put them all together although I have already mentioned mice
and woodpeckers. I have only selected a few below, varying
from the most serious to the less serious but common.

Foul Brood
There are two of these, American Foul Brood (AFB) and
European Foul Brood (EFB), which are unrelated. They are
both fairly widespread throughout the world and have no
geographical significance. In the UK they are both subject to
statutory controls; although confusingly for the beekeeper
each country has its own order. As soon as you start
beekeeping I suggest you find out the situation in your area
and who to contact if the need arises. This information is
probably best sourced from your local BKA. In some
countries Bee Inspectors are employed to inspect colonies and
would rather be called out to false alarms than have cases go
undetected, so beginners should never be hesitant about
reporting cases that they are not sure of.

AFB shows in the sealed brood stage, EFB in the unsealed
stage. Both of these diseases are uncommon which is their
danger, as many beekeepers will never see them. For that
reason it's important to know what healthy brood looks like
and if you see a problem to check it by looking at photographs
or calling in help. AFB is unmistakeable, but EFB can
sometimes look like other diseases.

I saw AFB in colonies within a year of starting beekeeping
and that helped me identify it when my bees were infected a
year or so later, but it was 43 years after starting beekeeping
that I first saw EFB in a colony and only because I asked the
Bee Inspector to let me have a look the next time there was an
outbreak locally. It happened to be in the teaching apiary of a
BKA where they were very strict on apiary hygiene, showing
that anyone can get it.

Foul brood can be introduced into an apiary in a number of ways including:

- On bees and equipment, e.g. purchase, allowing another beekeeper to put bees in your apiary, taking in combs for things like queen rearing or helping a colony out such as uniting.
- Robbing out other colonies.
- Much foreign honey is known to be contaminated because many countries have a much higher level of foul brood than we do, with far less stringent controls. Exposing your bees to any honey of unknown origin is bad practice.

For these reasons every apiary is at risk and early detection is very important. You may do everything possible to minimise foul brood, but still get it due to surrounding beekeepers not being vigilant. After introduction the spread of foul brood is mainly as a result of the movement of infected combs and equipment by the beekeeper.

Varroa

The arrival of the parasitic mite *Varroa destructor* has changed beekeeping forever, so beekeeping is far more complicated and difficult than when I started. It is here to stay and may never be eradicated, as even with the most effi-cient treatments there is a residual mite population in every colony. At the time of writing there are still a few isolated areas it hasn't yet reached, so if you are thinking of starting beekeeping please check locally to see if they have varroa, as in some of these places beekeeping may become impos-sible if it is introduced, adding another burden to the harsh conditions bees already have to deal with. Don't take the selfish attitude that it will eventually get there anyway, so it doesn't matter if you get bees from outside the varroa-free area.

The mite parasitizes both brood and adult bees by puncturing the body wall and feeding on the haemolymph ("bee blood"). They vector several viruses that were present in colonies at low levels pre varroa. The mites infect the larvae when feeding on them – the adult bees die from the virus infection. It is thought that more colonies die of the viruses than they do from varroa. Many of these viruses aren't detectable by the beekeeper, but one is, Deformed Wing Virus (DWV), where the emerging bee has little or no wing and as it obviously can't fly it is of no use to the colony at all. If bees are seen in a colony with DWV there is a high varroa level and urgent treatment is needed.

There are many things a beekeeper can do to reduce the level of varroa and you should seek up-to-date guidance elsewhere. These will include what are termed Integrated Pest Management (IPM) techniques and although not highly effective on their own they can be if used as part of an overall programme. It is very important that you fully understand the life cycle of varroa, so you know how the treatments work and if they are successful.

Chalk Brood

This is a fungus disease that is likely to occur in any colony at any time, especially in damp conditions. In my experience the use of OMFs and through ventilation in the winter are a great help as the hive is kept drier. The infected brood looks like little pieces of chalk, hence its name and in the early stages it can often look like EFB, so be careful. The bees can remove the diseased brood and it can often be seen on the floorboard or on the alighting board. Although common it is only a minor disease and usually won't kill a colony.

Nosema

There are two species, *Nosema apis* that has parasitized honey bees for a long time and *Nosema ceranae* that has jumped species relatively recently. There are differences in them but for the purposes of this book they can be considered together. They are a single celled fungal parasite that infests the gut of the bee and inhibits the digestion of pollen. Good husbandry will reduce nosema to manageable levels, but medication is possible. Some bees are much more susceptible than others and this can be dealt with when assessing your colonies.

Acarine
This is a parasitic mite that enters the trachea of worker bees. It is really only a minor disease in the UK and Ireland, although beekeepers should understand it.

Wax Moth
There are two of these, greater wax moth and lesser wax moth that are unrelated. It is the larval stages that do the damage by destroying combs, mainly those that have had brood in.

Lesser wax moth larvae are often said to only be a problem with comb that has had brood in, but on many occasions I have seen badly damaged super combs especially if they have been in store for some time. It takes some time for them to destroy combs and I rotate my supers, so what I didn't use this year I will use first next year and this is adequate.

Greater wax moth is much more serious and larvae will attack combs that have had brood in where it is very destructive. In the hive they tunnel under the brood cappings and leave what looks like a white silky line on the face of the sealed brood. Often the affected brood has its cappings removed, leaving the heads of the pupae exposed. This is known as "bald brood" and does not result in the death of the pupae. Outside the hive the speed with which they work is temperature-related, as they can multiply very quickly in the warm weather. They chew the combs and produce solid patches of cocoons, reducing good combs to a matted mess that is irrecoverable in a short time. They can chew wood and damage the inside of brood boxes if the frames are packed together. I have always used natural methods in dealing with them and with care there is no need to resort to anything else. It is a case of constant care and attention:

- Stored brood combs will be a problem, especially if packed close together, so leave them on the normal spacing. If you no longer need them either melt them down or dispose of them.
- Freezing kills all stages; a week in the freezer works well.
- If you keep your colonies strong you shouldn't have problems inside the hive. Some races of bees are said to be

more tolerant, but I have never seen a bad case in a colony
of mine.

Wasps

Wasps are usually only a problem in late summer when their
needs change from protein to carbohydrate. They are much
tougher than bees and on a one to one situation a wasp will
always win. They are looking to gain entry into a colony to
take honey and they aren't usually a problem if colonies are
kept strong and entrances closed down so they can be easily
defended. It is probably little use putting out wasp traps
because this is the time of the year when wasp nests are at
their strongest and there are usually many nests within flying
distance of a colony of bees.

Other Pests and Diseases

There are several others, but in general good management
practices will keep them to a minimum and there is plenty of
good information available to advise you. We must be aware
of all pests and diseases and know about them. Some can only
be diagnosed by microscopic examination or by other means
such as is needed for viruses. There are varying attitudes to
disease and the extremes are paranoia and total dismissal,
neither of which is any help to the rest of us. The competent
and responsible beekeeper will:

- Learn.
- Understand.
- Recognise.
- Treat.
- Be careful.
- Stay alert and vigilant.
- Keep learning.

This can be encouraged by local BKAs who will regularly
have presentations and demonstrations, possibly in conjunc-
tion with the Bee Inspection Service if there is one, invest in
microscopes and train members to use them. Disease levels
are determined by all of us and BKAs have a big role to play

in education, but they need co-operation and willingness to learn from beekeepers.

Robbing

Robbing is a term that is used to describe when a colony is under attack from other bees or wasps who are trying to steal their stores. It is quite easy to spot because there will be more activity than normal at the entrance and possibly fighting as well, although when in an advanced state this may not be seen, as the colony being robbed may lose the will to defend itself.

Continual observation at hive entrances will give you an idea when things aren't right. Bees that are returning to their own hive will land at the periphery of their entrance then walk in. Robbers usually go for the centre after flying in darting movements.

If you smoke the entrance and take the crown board off, any robbers will immediately fly out, not stay calm as a normal colony would.

Wasps are much tougher than bees. They can fly at lower temperatures and in a one to one situation are likely to overpower a bee. Quite often you will see wasps entering a hive at dusk when bees have stopped flying.

I think it's fair to say that if a colony is being robbed it is usually a beekeeper-induced problem, or certainly on the few occasions I've had it I considered it was my fault, because I was careless. Prevention is better than cure. It is most likely to happen when there is little or no nectar coming in and if a colony is picked on, robbing is quite difficult to stop. If you catch robbing in the early stages you have a reasonable chance of stopping it, but once it takes hold stronger action is needed. If you shut off one robbing opportunity you could transfer it elsewhere, so you must be watchful.

Make sure your colonies are strong, entrances are closed down to a single bee space so the bees can easily defend them, there are no gaps in the hive for bees or wasps to gain access, no honey or syrup is left exposed and that you feed at dusk. Colonies can lose morale if they are queenless for some time, so try to keep them queenright.

If you do have a colony being robbed out I suggest you close it up quickly and move it more than 3 miles away.

If you have left something like honey or syrup for bees and wasps to access you should remove it quickly and replace it with a small amount of the same. When the robbers have cleared it up they will be satisfied and should calm down. If you simply take the source away they will search for it for some time and may pick on a weak colony instead.

Although we can do a lot to prevent our own colonies being robbed we are unable to stop them robbing other people's. This is one of the major ways diseases are spread, particularly foul brood and nosema. For this reason we must check our colonies regularly, understand diseases and be able to recognise and deal with them.

Apiary Hygiene

This should be a common sense matter with the individual beekeeper having his or her own regime. As already stated many diseases will come into your apiary on bees that have been elsewhere, so vigilance is required at all times. Robbing is one of the main causes, but bees, especially drones, visit other colonies and it's reckoned that a fair number of bees in a colony didn't emerge there. You have little or no control over this.

Another common way disease is transferred is on bees and equipment coming into the apiary. Here the beekeeper does have control and some suggestions to minimise problems are:

- Any bees can have a brood inspection before entering the apiary.
- Used equipment that has come from an unknown source can be scorched out or burned. Unused combs that have had brood in are probably safest burned; those that haven't had brood in can either be melted to recover the wax or treated with acetic acid.
- Hives can have propolis and wax scraped off, then scorched out with a blow torch.
- Don't allow anyone else's hive tools to be used. Gloves should be clean or sterilised in washing soda solution.

For the use of acetic acid and washing soda I refer you to the excellent leaflets produced by the BBKA.

Probably the most likely diseases to be spread in this way are foul brood and nosema. I think the best approach is to assume there is a chance of all colonies in the apiary being at threat of all diseases all the time. This needs some sensible procedures, and the understanding and recognition of these diseases are important aspects.

It is easy to check for both foul broods at every colony inspection, but nosema will need microscopic examination.

There is a modern view that I have seen in print and heard from lecturers that if you sterilise your equipment between inspecting each hive you won't get or spread foul brood, but I think this is a wrong and very dangerous approach. This gives the impression to beekeepers that you don't have to look for disease, just sterilise everything and you won't get it. Now let's think about this logically and bear in mind that your bees can easily be infected without your help. Sterilisation is only likely to be effective if you have foul brood in the first place, so why not look for it? I always encourage people to check one frame with brood in **all stages** in **every** colony at **every** inspection, the unsealed brood for EFB and sealed brood for AFB. I think that is a much more sensible approach and it should be a habit, takes no time at all, aids your learning and awareness of what's happening in the colony and is likely to detect the disease at an early stage. There is the added benefit that you will learn about other diseases as well. Sterilising hive tools, smokers and protective equipment together with good apiary hygiene will certainly focus the mind on disease, but they probably won't prevent its spread if you already have it and don't notice it, so I repeat that in my view it would be much better to know what healthy brood looks like and to learn about diseases and how to recognise and deal with them.

13

RECORDS

For the first year or so of beekeeping I wrote an essay on each colony at each inspection, but I soon asked myself what benefit it was. It took me longer to write about it than to inspect the colony, so I gave up. When I was young enough to have a good memory and we didn't have the current problems I didn't bother with records, even when I had 130 colonies, but I do now and I suggest everyone else does as well. My knowledge of my colonies is much better, even though I only record minimal information.

There are many ways of keeping records including an exercise book and a standard record sheet that may be from a BKA or found on the web. In my opinion the vast majority are far too complicated and I once saw one with 32 entries you could fill in on every inspection! In my view beekeeping is a hobby that should be enjoyed, not a chore. Many record trivial things and you wonder why you need to know some of it anyway.

I designed mine on the computer and I record simple meaningful things that are relevant to the next visit, yet I can easily see the colony history. It is an A4 sheet for a colony for a year and kept in a plastic wallet in a clip board under the hive roof. It has the information shown overleaf:

- **Date ../../..**

- **Queen Laying Y/N**
I don't need to know any more.

- **Queen Clipped & Marked Y/N**
This tells me if I have seen the queen or not and if she is clipped and marked. If I don't see her I leave it blank.

- **Queen Cells Present Y/N**
This is good information for the next inspection.

- **Temper**
I mark out of 10. This tells me a lot about which queens I cull and which I breed from.

- **Calmness on Comb**
Same as "temper".

- **Comments**
This is where I note such things as chalk brood, moved brood frames, where an introduced queen came from, etc.

The Y/N – for "Yes" and "No" – are crossed out and the whole sheet is very easy to read. I have used these sheets with the occasional modification for several years and they tell me all I need to know. It is also a simple matter to keep them.

   This record sheet suits me and it is a simple job to add the things you want to record.

14

FOR THE FUTURE

In beekeeping there is always more to learn and more to do. I have tried to give you an idea of the absolute minimum you will need to get you going, but everybody has different aspirations, ability, time and rates of learning. Years of experience in teaching beekeeping has told me the best approach is to give enough information so the average person can understand the principles and those who learn quickly have the opportunity to go as far as they wish without being held up. For that reason I give here a few ideas of some subjects you may find useful after you have settled in and feel a need to move on.

Queen Rearing and Bee Improvement

In general most bees are very variable and quite ordinary. There are some good ones, but there are also some dreadful ones. I find it amazing that many beekeepers are prepared to tolerate bees that have poor characteristics such as bad temper, running on the comb, bad wintering, susceptible to disease, etc. These can quite easily be improved by assessing colonies at every inspection, rearing queens from the better colonies and culling the worst. A lot of beekeepers think that queen rearing is difficult, but in fact it is quite easy. In simple terms all you have to do is to encourage the bees to convert worker larvae into queens and we all have many opportunities during the summer where the bees present us with that opportunity. Bees that are productive and easy to handle are a joy to work with.

Queen Introduction

I have mentioned requeening a few times and you may need to do this early on. For the needs of a beginner I will keep it simple and assume you need to replace a laying queen with another laying queen. As you are likely to get a queen from another beekeeper I think you should seek help or guidance from him or her the first time you do it.

The queen substance produced by all queens is individual and instantly recognised by the workers, so if you tried to put a different queen in a colony the bees would pounce on her and kill her. To avoid that we remove the existing queen and replace her with another one that is in a cage. Queen cages are available from appliance dealers, there are several types available and I think the best for a beginner is the plastic one that's called a "puzzle" cage. It has a sliding lid making it easy to get the queen in, holes for the bees to feed the queen and somewhere for the queen to hide to prevent the bees biting her legs which can happen soon after introduction. See Fig. 9.

Insert the cage between frames in the brood area, or lay it on the top bars of the frames, but make sure the bees can feed the queen. The queen substance from the old queen will diminish, the bees will realise there is one in the cage and start feeding her. In doing so they take queen substance from her

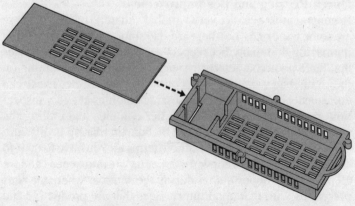

Fig. 9   Queen cage, puzzle type

which is dispersed throughout the hive in the normal way and after a time she is accepted as theirs. After 24–48 hours the queen can be released and the operation should be successful, although it can never be relied on to be 100 per cent and is one of the things in beekeeping that is getting more difficult.

Queen Performance

In the first few years of the twenty-first century I have been trying to highlight problems many people both here and abroad have been experiencing with queens. These are very early supersedure of young queens, early queen failure and queens simply "disappearing". There also seems to be increasing difficulties in getting colonies to accept introduced queens. None of these problems should happen at the high rate they are. I have tried to encourage research, but there are many other things that need resources in times of cutbacks and it is often difficult to get people to recognise there are problems even though they are widespread. I believe the reasons are very complex. I have written articles for magazines and websites and there are many references by others to the issue. For these reasons bees may not always do what beekeepers expect.

Harvesting Beeswax

Sadly there is much beeswax wasted, which is a pity considering the cost to the bees to produce it. It can be melted in a solar wax extractor that is free to run. Although they are quite expensive to buy they are quite simple to make for a practical person. Wax is needed to make foundation and can be exchanged, sold or traded in for equipment. It can also be made into soap, cosmetics, candles, etc.

Microscopy

A relatively cheap microscope can be used to identify some common diseases. A good BKA should have at least one of their own and members available who can train others in its use. A step further could be the study of anatomy and pollen analysis.

Uniting

Uniting (see Fig. 10), which can be carried out for a number of reasons, is when two colonies are put together. The term that everyone understands is "uniting", not "amalgamating", "joining", "merging" or "combining" as we see in some places.

At any time of the year and particularly towards the end of the season, there is often a need to unite colonies, usually to turn two or more weak ones into one stronger one that will have a better chance of survival. A good beekeeper will realise why the colonies are small. This could be disease in which case you need to deal with it. Uniting can also be done to reduce the number of colonies. Allowing for different situations and assuming disease isn't a factor I unite as follows:

- Bring the two colonies together side by side. This can be at the rate of 3 feet a day in good flying weather.
- If both colonies have queens, kill the worst one.
- I prefer to have the stronger colony at the bottom.
- Move one hive to one side.
- Move the other hive midway between the two hives, remove the crown board and place 2–3 sheets of newspaper over the brood box. Some sources suggest one, but I have always found 2–3 works well.
- Either make a couple of slits through the newspaper with a hive tool or pen knife, or smear some honey on a small patch on opposite sides. This needs to be done over the gap between the lower frames.
- Place the other brood box on top.
- In my experience it doesn't matter if the remaining queen is in the top or bottom.
- Close down and leave for a week.
- A week later rearrange combs to make a reasonable brood nest.

Uniting is best done in the evening. I usually try to keep my colonies strong, but there are occasions when uniting is needed, such as at the end of the season when you have a poor queen or a queenless colony.

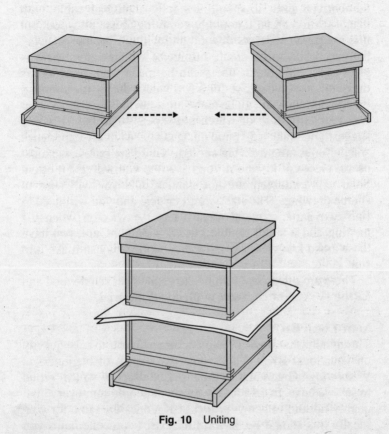

Fig. 10   Uniting

Making Equipment

If someone is handy with tools there are several things that can be made quite easily. Any hive parts need to be correct and bee spaces observed. There are drawings on several websites that can be downloaded and they should be adhered to. Don't copy existing parts as there is a manufacturing tolerance and wood shrinks. In general the inside dimensions and depths are critical and you should fully understand the workings and needs of a hive before attempting to make one. If you stick to the correct sizes it should be fully interchangeable with manufactured parts. It could be tempting to use plywood, but it is not porous like solid wood and will not breathe, causing the brood box to be damp during the winter unless there is plenty of ventilation. Supers can be made from plywood as they are only on the hive for a few weeks during the summer.

Other things that can be made are a nucleus box and a solar wax extractor, both of which are expensive to buy. I would suggest you make a 5-frame nucleus box, as in my experience smaller ones are never big enough. These are one of the most useful pieces of kit and if you have one you will use it often. Solar wax extractors are not standard and you will need to source drawings. The sizes aren't critical and you will need to find two parts: a metal tray to rest the wax on when it's melting and a small double glazed window. Once you have these and a knowledge of how they work, you can make it to suit.

There are often opportunities to acquire recycled wood and a keen eye can often result in plenty of material.

Your Local BKA
Throughout this book I have encouraged you to rely heavily on your local BKA. That's what we expect of beginners at Wisborough Green and in the last five years or so I'm proud to say we have given a lovely group of people some very high quality tuition. I hope your local BKA does the same for you. Ideally this should be on a rolling basis: someone helps you and when you are experienced enough you help someone else. Please don't just grab what's on offer then disappear into the distance when you think you have learnt enough.

Bee Fever
I have left this to near the end in case you have forgotten it by now, but I must warn you as it's quite common. Bee fever only comes in one size XXL and I've seen it many times. I am well qualified to write about it as I put 24 colonies into my second winter, but only half of them survived. Why did I lose them? Several reasons including:

- Not putting on mouse guards, as very few did in those days.
- Expecting small colonies to come through the winter. I wanted to increase my colonies quickly.
- Using Italian queens that kept breeding and used up food much quicker than I expected, so they starved.
- Possible nosema due to using bees that were susceptible to it.
- Old hives with leaking roofs so the bees got wet. Much of the equipment in use in my locality was very old and I acquired the junk others threw out.

Did I have any advice? No, I didn't. Did I learn? Of course I did and fast. I wouldn't make any of those mistakes now and I don't think any of my local BKA members would either. I had nobody to warn me, but you and they have.

I always advise beginners to get several meetings under their belt and get some early experience at our teaching apiary. Then we will try to help them source bees in the knowledge they are streets ahead of many others and with a bit of guidance they can have a second colony quite quickly. They should experience a couple of winters, then expand gradually if they wish. This could save everyone a lot of bother as with this approach we now find a lot less give up after a couple of years or so than have in the past.

15

BEGINNERS' TIMETABLE

This section has been modified from a guidance sheet I wrote for beginners of Wisborough Green BKA and it has been on our website for some time.

I'm assuming you will have the opportunity to handle bees under the supervision of an experienced beekeeper who is teaching you good handling skills, the basics and the theory to a reasonable degree. Everything I have set out below is factual and is relevant to whatever methods are taught.

There is absolutely no use inspecting a colony if you don't know what you or the bees are trying to do.

These notes are intended to help the beginner in the early stages and I have devised the following suggested timetable that will give them something to aim at, although people, bees and situations can vary considerably. It should be seen as a minimum requirement. There are other things you or your tutor may wish to add.

Some of the following may take place before you have your own bees:

First Colony Inspection
Ideally this should just be to get you handling a colony entirely on your own after being shown what to do. There is a lot happening in front of you, it is all new so don't try to take too much in. Try to handle the frames gently, but don't worry too much about what is on or in them. If you have any fear,

then the person supervising you should detect this and help you. If you are genuinely frightened, tell your tutor, as there is no point carrying on. Don't worry how long things take: this is the one colony inspection you will remember for the rest of your life. The most important thing is to be a confident and competent bee handler and of course to enjoy it.

Second Colony Inspection
Be aware of the three castes, i.e. queen, drone and worker. Their roles can either be learnt from your tutor, or other sources, but the sooner you learn them the sooner you will understand many things that happen in a colony.

Be able to see eggs and young larvae. If you have a problem then sort it out at this stage, as there is no point struggling.

Now you are feeling more comfortable, your handling techniques will be developing. Many things you do will become habit, e.g. "reading" the colony and smoking when needed, removing frames without "rolling" bees, noticing the queen without necessarily looking for her, seeing eggs, making sure the colony has enough food, looking for disease, holding frames above the brood box until you know where the queen is, standing up straight so you don't get backache, assessing the colony, etc, etc. If these things become a habit early on, you won't have to worry about them later.

Third Colony Inspection
Be able to recognise and name the main hive parts and know their uses, i.e. floorboard, brood box, queen excluder, super, crown board, clearer board, roof, frames and foundation.

Light a smoker and keep it alight. This is one area where many people have problems, but is absolutely crucial especially if a smoker goes out part way through inspecting a "touchy" colony. Keep practising away from the bees if you want to. Once you have mastered it, keep a constant lookout for fuel that suits you.

Recognise pollen and liquid stores and know their uses and importance to the colony.

Find a queen without help.

Fourth Colony Inspection Onwards

Be able to recognise queen, drone and worker cells and brood in all stages. The earlier you can assess their ages, the better.

Learn off by heart the life cycles of each caste and the swarming procedure. This is most important and will help you assess the state of a colony and help you address many of the problems that might arise. It is absolutely crucial in any method of swarm control.

Recognise healthy brood in all stages. If it doesn't look right, there is probably something wrong with it. Refer to photographs that are readily available. Do not get paranoid about diseases, but a glance at a comb or two in every colony should be part of your normal inspection and takes no time at all.

Understand why 7 or 14 day colony inspections are needed.

Be aware of varroa, be able to see the mites and the signs of heavy infestation.

Recognise the three types of queen cells based on their shape and quantity.

You should now have enough knowledge to open a colony on your own and understand its workings. The knowledge gained so far should help you to progress further.

The following should be known if you have your own bees:

Before August

Learn and understand the life cycle of the varroa mite as this will help you understand treatment methods. Learn about the current treatments that are available and be aware of IPM techniques and the reasons for their use. Learn how to monitor mite levels as this has become a very important part of beekeeping. It is important that you know the levels of mite infestation both before and after treatment. Varroa and the viruses it vectors have probably become the biggest killers of colonies and it would be disappointing for a beginner to lose a colony when simple measures would have saved it.

Most beginners will only be putting one colony into their first winter, so to avoid disappointment read Chapter 11, Wintering.

Of course some of the above can be learnt before you have

your own bees and you should take any chance you can to
handle a colony. Although I wouldn't advocate getting stung
deliberately it wouldn't be a disaster if you did. It is surprising
how many people take up beekeeping with great gusto and
give up quickly because they don't like getting stung.

End of the Active Season
By the end of the first season you should as a minimum be
able to open a colony without help and handle it with a reasonable degree of competence. This will give you the basis on
which to further your knowledge during the winter, ready for
your first full season in the spring. I suggest you concentrate
on the following subjects:

* Queen substance. This is a pheromone which should be
  seen simply as a chemical stimulus. There is no need to go
  into any great detail, just try to understand the influence it
  has.
* Study honey. It is hygroscopic, find out the relevance and
  possible results.
* If you intend selling honey be aware of the regulations.
* Investigate the uses for your hive products. There are many
  books on such things as cooking, honey drinks, candle
  making, polish making, etc. If you haven't had a crop
  already, you should have next year.
* Plan sensibly for the coming season, but have a chat to
  others to get ideas, as they have all been through it as well.
* Learn the principles of queen rearing and bee improvement.
  The quality of bees can be rapidly improved without any
  great knowledge or equipment. Your bees may well present
  you with opportunities during your second season. Speak to
  someone who regularly rears queens and beware of those
  who "just let the bees get on with it".

Provided that you had sound tuition to begin with, have
continued to attend demonstrations, learnt at a reasonable
speed, learnt how to handle a colony with confidence and to
recognise its condition, deal with the more common situations
and are willing to learn more, you should no longer be classed

as a beginner, but an improver and firmly in the intermediate group.

By the end of the second season many people are actually very knowledgeable and competent and will be able to:

- Dispense with gloves.
- Dispense with bee suit so that you are only wearing a veil or tunic.
- Find a queen.
- Be able to clip and mark a queen.
- Recognise the signs of EFB, AFB, Varroa, Braula, Nosema, Acarine and Chalk Brood, but bear in mind you may never see some of these.
- Clear supers, extract and process your own honey.

If you have got this far, well done. I hope you enjoy the craft and I ask you to try to encourage and help others and remember:

Don't do anything unless you understand clearly what you are trying to achieve, what the end result should be and have a "Get out of jail" card up your sleeve. In beekeeping you often need it.

16

CREATING A BEE-FRIENDLY GARDEN

Most beekeepers keep their bees at home and are gardeners, or if not someone else in the family is. This gives an opportunity to plant species that will attract bees and increase the enjoyment of beekeeping.

One of the reasons often given for starting beekeeping is to improve the cropping or number of flowers produced on plants in the garden. The latter shows a basic misunderstanding of pollination where surprisingly a significant number of people seem to think that plants will bear more flowers as a result of being pollinated by insects. Pollination is the act of transferring pollen from the male part to the female part of a flower either on the same or a different plant, resulting in fertilization that produces seed which may be an ornamental or edible fruit, or just a simple seed for producing more plants.

The keeping of bees in a normal garden won't increase the yield of your plants as there will usually be sufficient pollinating insects anyway. The planting of plants that are attractive to bees in a single garden won't increase the honey yield of a colony of bees as they will normally forage within a radius of perhaps 1½ miles, so whatever can be planted will be insignificant on its own. However, as a colony of bees will provide a large number of pollinators for a wide area, if more people planted plants that provided food for insects, not only would it provide a good overall source of forage for bees, but also for other pollination insects which should increase their numbers in the locality.

Although this book is about beekeeping we should not forget that what suits bees will often suit wildlife in general and perhaps instead of thinking rigidly about a bee-friendly garden we should think about a wildlife-friendly garden with an emphasis on bees. Gardens with flower beds and borders that have a blaze of colourful double flowers, double flowering trees and shrubs and lawns closely mown with no plants in them at all are almost useless for wildlife, especially if they are regularly sprayed with herbicides and pesticides in order to keep them "looking good". In general, double flowers are no use to bees, but single ones are.

There are over 250 species of bees in England, Wales and Scotland, rather less in Ireland. The vast majority are bumble bees or solitary bees and although many species of plants will suit many species of bees there are several reasons why you won't see all types of bees on all types of plants. This could be for such reasons as the adaptability of flowers to suit some bees, or in the case of many solitary bees when the adults are present, many of them having short adult lives. Honey bees are the only ones that live as a colony throughout the winter and provided that the weather is warm enough for them to fly they are generally able to do so throughout the year. As far as foraging is concerned honey bees are unlikely to be seen on flowers between the end of November and the end of January, although there will always be exceptions, especially in the warmer areas.

Gardening has become popular and this has led to the increase in the number of garden centres. In general I think they do a good job, but they largely sell the "standard" range of plants and seeds. There are many specialist suppliers that are well worth investigating. Although you may pay a shade more with them they will have available a vastly increased number of species and will provide many interesting plants that more mainstream outlets may not have. The good thing about these suppliers is you can often get good sound advice and you may be able to visit them to see the plants growing. With some plants, soil types and conditions are important and it may be helpful to seek expert advice rather than read the minimal information on the back of a label.

If you take your time planning and selecting what you want rather than load up a trolley with what is instantly available you may end up with a better result and you may be happier with it. It would be a good idea to visit a supplier every few weeks to see what is in bloom. This would help in developing your garden and spread the cost. I won't mention many species here but what I do will only be in general terms and what is commonly available.

We tend to think of plants that are useful to bees as having to be big and bright, but this is not the case, as often plants with fairly insignificant flowers are quite useful in a number of ways. A good example is *Cotoneaster horizontalis* that feeds a number of insects in the spring, has a mass of red berries providing colour in the autumn and a useful food source for birds in the winter and is good for covering up walls and fences. Unless totally neglected it never looks untidy.

Escallonia is in my opinion a much underrated plant and seen as old-fashioned. It can be used as a specimen shrub or as a hedge. It grows almost anywhere with some varieties being quick growing. The flowers vary from white to red, but mostly pink and appear from mid-summer to autumn. It has shiny evergreen leaves so looks good all year round and can be kept neatly clipped.

Some plants can be used for a dual purpose such as top fruits like apples, pears and plums where you get blossom and fruit as well and possibly shade if you need it. Apples and pears can easily suit a fairly small garden if they are grafted onto dwarfing rootstock where they can be trained to take several forms of which the cordon is very useful. This does need a reasonable degree of knowledge about pruning otherwise they may become rampant or the fruit buds cut out. With top fruits I suggest purchasing from a specialist grower where you should get good advice and a much wider variety of cultivars than is normally available. Make sure you like the fruit as there is little point buying an apple that is sharp with firm flesh if you prefer them soft and sweet. The range of fruit sold in supermarkets is more restricted than that formerly available which means that many people just think of apples being Bramley or Cox. However, the former produces large trees and are tip bearers making the pruning of small trees difficult, while the latter is very susceptible to canker on some soils, so neither of these varieties is ideal in a small amateur garden. Nevertheless, there are substitutes that a good grower would suggest. Some varieties are biennial bearers; others may need suitable varieties as pollinators, so carry out a little research.

If there is space in your garden, soft fruits such as raspberries, gooseberries and blackcurrants are well worth growing and they all attract insects. They are all productive and will freeze well. A fruit cage will probably be needed but that may be seen as unsightly by some.

Borders and beds can be planted with a variety of things such as shrubs, perennials and annuals. These are the areas of the garden where there is likely to be most colour and where it is expected to see the most wildlife including bees, simply because the eye is taken to it. As well as consulting books it would pay to keep your eyes open when visiting other gardens for plants that have bees on. You may be able to help with a little dead heading as well! A packet of mixed annual seed or several packets of the types you want and mixed will go a long way if broadcast, although you may need to deal with slugs, as they will enjoy young fleshy seedlings. There is little effort needed here apart from rough digging in the autumn, raking over in the spring and sowing again. Weeds are rarely a problem and you could have a mass of colour within a few weeks of sowing right through to the frosts. I suggest you include single poppies as they provide bees with a large amount of pollen and their seed heads can be attractive. Night scented stock isn't a great bee plant, but the scent is so good, especially on warm still summer evenings that I encourage you to include it in your mix. If you collect seed heads from the types you like or are attractive to bees you can use these the following year. All you need do is keep them dry and away from mice.

One perennial plant that is very attractive to a large number of insects in late summer and autumn is *Sedum spectabile*. There are several varieties, it is easy to grow and suppresses weeds, with attractive seedheads that provide interest in the winter. Michaelmas daisies are seen as old-fashioned plants but are useful at the back of a border and very easy to grow, although the taller varieties may need staking. Lavender is a favourite of most people and very versatile with a large number of varieties and habits. The seed heads can be harvested and most people can find a use for them.

Spring bulbs such as snowdrops, crocuses and single tulips all provide useful sources of pollen as well as colour and can be grown in containers. Snowdrops are better bought "in the green" rather than as dry bulbs.

Confined or paved areas need special treatment as containers may be needed. Selection of plants will need to be done carefully and so will maintenance, with watering being a particular problem. A container on a sunny patio may look a bit thirsty after two weeks' holiday, so friends and neighbours may need to water for you. There are many suitable plants that bees like and marigolds are very useful, but remember to dead head them regularly to keep them flowering, otherwise when they have set seed they won't continue to flower so well. They are easy to raise from seed and you will always find the odd seed head that can be dried and used for the following year, so the cost is next to nothing. Containers will probably need feeding, but check the feed as one high in nitrogen will produce leaves, where potash will encourage flowering.

Bees stay loyal to one flower type during each foraging expedition so they won't mix species. This is quite handy from a plant's point of view otherwise they will be getting pollen from another species that won't be any good to them. I have come across several people who thought that planting flowers near their fruit trees would encourage bees to pollinate their fruit, but that doesn't happen.

For those with larger gardens there is an excuse to be a little untidy as many wild flowers are very good for bees, for example, dandelions. The ivy that is growing over the fence or up a tree will provide nectar and pollen for a number of species in the autumn and on a warm day it can be humming with activity, so don't cut it down just to be tidy. Contrary to popular opinion ivy doesn't kill trees. It may look that way, but it is probably taking advantage of the fact that the tree is on its way out anyway so there will be fewer leaves and live branches to inhibit it.

In a more formal garden trees will be important, but in general they will take some years to establish. Limes are quite useful, but unless existing they are probably not worth planting. Pussy willow (*Salix caprea*) is a very useful source of spring nectar and pollen, but it is dioecious, meaning that the trees are either male or female. They do strike readily from cuttings so are easy to propagate.

Lawns with a few weeds in such as white clover that are not cut short will provide food for bees and any odd grassed areas such as orchards can be left uncut until the autumn to allow wild plants to seed. The regular mowing of grassed areas simply to be "tidy" does little to attract wildlife of any kind, when if left could provide a wonderful display of summer flowers and food for insects as well as a much richer habitat for other wildlife.

As well as providing useful food sources for bees and other insects, careful planting will provide much interest for the keen observer. This can lead to other things such as the search for further knowledge or to give you an opportunity for a bit of photography or painting.

Even if you don't take up beekeeping it would be nice to know you gave a bit of thought to other creatures when planning your garden.

As well as providing food for bees we also need to think about habitat and that is often overlooked. Bees need to nest somewhere and some bees are quite specialised. Honey bees need a cavity that is about the right size, dry, protected from pests such as mice and with an entrance that is easy to defend. Their natural home is most likely to be a hollow tree but it's only the mature garden that will provide this. As an alternative they will use buildings, especially chimneys, cavity walls and behind tiles. In reality the honey bees in your garden are likely to come from a hive belonging to a beekeepeer and it is impractical to do anything yourself to stop them building such nests apart from taking up beekeeping.

If you do have a nest of honey bees they are unlikely to do any harm unless the entrance to their nest is in an inconvenient place such as by a door or window, or at head height where they may be a nuisance to those walking by. My advice usually is to leave a nest if possible because swarms of bees like to go where bees have been before and if they are removed they are often soon replaced by others. You will get a swarm occasionally, but I have known of many similar instances and they are usually no trouble.

Bumblebees and solitary bees will nest in a variety of places, often in the more untidy part of the garden. Many bumblebees like to nest in the ground where they use old mouse or vole nests, although they will use many similar places higher up such as tit boxes that still have the nest in from a previous year. They prefer a degree of shelter and this can be inside or under a shed, or a great favourite is a compost heap or bin.

Solitary bees are very diverse in their choice of a nesting site and these can include holes in dry ground, banks, walls and in wood that may have been made by boring insects. It is possible to buy nests for bumblebees and there are many designs available in books or on the web. Nests for solitary bees can be easily made by using a piece of plastic pipe or a piece cut out of a plastic bottle about 150mm long. Cut off some pieces of bamboo cane a little longer, fill the tube with them and hang it on an outside wall. An alternative is to drill holes in a piece of wood varying from 6–10mm diameter about 60–80mm deep, but angled upwards slightly so the rain doesn't fill the holes with water. This can also be fixed to a wall or shed. These nests need to be in place before the bees are looking for a home in spring.

17

CONCLUSION

For those with the ability to understand what happens in a colony of bees, can recognise problems and deal with them I think beekeeping is a wonderful hobby that will give endless pleasure for many years. I promise you there is plenty to learn and I find that one of the many attractions.

I warned you that in places I would express strong views and I hope I haven't overdone it. Please accept that I merely want bees to be treated in a respectful way, by people who understand them, or are willing to do so. Some of the things that have crept into beekeeping in recent years annoy me as they have often been dreamt up by those with little or no knowledge and repeated by others so often they have become "fact".

I have got this far without consulting any other book and virtually everything is exactly as I have done it or teach it and being well known in my area I wouldn't be allowed to get away with it if it was otherwise. Some things I advocate may not appeal to everyone, but when put together they form a system that should work in all areas of the country. As I have already stated, some of my advice is specifically for beginners and I wouldn't always offer it to more experienced people as a first option.

There are many different ways of keeping bees and I fully accept that. The main problems come when different methods are mixed, but taking advice largely from one source should avoid that. Be open minded enough to listen to others and make up your own mind what to do, but bear in mind what I said about prejudice and intolerance. Successful beekeepers will be able to devise their own methods from the many there are available.

In general I find beekeepers nice people who tend to help each other and I still enjoy teaching them as much now as I ever have. I am very lucky at Wisborough Green that we have some beekeepers with tremendous potential, some excellent demonstrators, including a 16-year-old we have just appointed and responsive members. There is a huge amount of banter at meetings, most of which seems to be at my expense. I hope that high standards and enjoyment are the main features of your BKA as well. Bees deserve it and I wish they could understand how much they contribute to my view that **beekeeping is fun!**

GLOSSARY

Acarine   A mite that enters the trachea of bees.

AFB   American Foul Brood. A serious brood disease that mainly shows up in sealed brood.

Apiary   A place where one or more colonies of bees are kept.

Bee Space   A gap that bees use for access in their nest. Their reluctance to fill this enables us to manipulate a hive by designing hives with a bee space between frames and the boxes they fit in.

Braula   *Braula coeca* is a wingless fly that can be mistaken for varroa mites. They are found on workers and queens, but are virtually harmless. Varroa treatments have controlled them, but they are still found in areas without varroa.

Brood   Covers all stages of development up to the emergence of the adult bee from the cell.

Brood Box   A topless and bottomless box, usually at the bottom of the hive where the brood is reared.

Brood Chamber   Another name for a brood box.

Capping   The seal on cells of honey to preserve it, or brood to pupate. Term also used for the act of doing it by the bees.

Cast   A second or subsequent swarm. Will contain one or more virgin queens.

Caste   The three different bees in a colony, i.e. queen, drone and worker.

Cell   The individual hexagonal sections that make up the comb.

Clearer Board   A beeproof board with a one-way escape mechanism to remove bees from supers prior to honey extraction. Often also used as a crown board but without the escape.

Clipping Queens The removal of one part of a wing or pair of wings of a queen to prevent her flying.

Colony   Usually taken to mean the whole population.

Comb Honey   Honey that is not extracted but eaten with the comb.

Crown Board   A flat board that covers the top of the hive underneath the roof. Often also used as a clearer board when fitted with an escape.

Drawn Comb   Structures of individual cells making up the nest. Made by bees with beeswax.

Drifting   Where returning bees go into other colonies, usually when blown by strong winds. In some parts of the world it means bees leaving honey supers that have been left in the open to clear them of bees.

Drone   The male honey bee.

EFB   European Foul Brood. A serious brood disease that mainly shows up in unsealed brood.

Emergency Cell   A queen cell that has been built on an existing worker larva in response to the queen going missing from the hive.

Floor   The lowest part of the hive. Can be solid or OMF.

Foul Brood   A term used to cover AFB and EFB.

Foundation   A flat sheet of beeswax embossed with the form of the bottom of cells. Used to give bees a start when drawing comb.

Frame   Used to fix comb in to make colony inspection easier. Normally wood, but occasionally plastic.

Hive Tool   A tool for prising apart hive parts.

Honey   The carbohydrate part of the bees' diet. Processed from nectar by the bees.

Honey Extractor   A device for removing honey from combs by centrifugal force.

Nectar   The sugary substance collected from plants that bees turn into honey.

Nosema   A term used to cover *N apis* and *N ceranae* which are adult bee diseases.

Nucleus   A small colony of bees usually on 3–5 brood frames.

OMF   Open Mesh Floor. A floor where the base is not solid but made from fine wire mesh.

Pheromone   A chemical that is secreted by an animal that influences the development or behaviour of another of the same species.

Pollen   A granular substance collected by bees that forms the protein element of their needs.

Pollen Basket   An area on the back legs of worker bees where they carry pollen and propolis back to the hive.

Pollen Pellets   The compressed pollen that is carried in the pollen basket.

Propolis   A sticky substance collected by bees mainly from trees. Carried back to the hive in pollen baskets.

Queen   Fully developed female whose main functions are to lay eggs and produce pheromones.

Queen Cage   A device for containing a queen, usually used to confine her when introducing her to another colony.

Queen Cell   The structure where a queen develops from egg to emerging adult.

Queen Cup   The base of a queen cell before it is extended into a queen cell.

Queen Excluder   A device with slots in that is placed between the brood box and supers that prevents the larger queen and drones from entering the supers.

Queen Introduction   The process of getting a colony to accept a queen that had not emerged in that colony.

Queen Substance   A pheromone produced by the queen.

Queenless   The state of a colony that has no queen of any kind.

Queenright   The state of a colony that has one or more queens that may or may not be laying.

Robbing   A term used when wasps or bees from other colonies are entering the hive and removing food stores.

Runners   Bees that are not calm on the comb. They appear agitated. Also the metal or plastic strips fixed to the brood boxes and supers for frames to rest on.

Skep   Straw or wicker beehive traditionally used by beekeepers. Modern skeps are used for collecting swarms.

Smoker   A device with a firebox and bellows. Smouldering fuel produces smoke that is blown at the bees to control them.

Splitting the Brood In a normal situation brood is on adjacent frames. If the beekeeper inserts a frame of comb with no brood on between two frames of brood it is termed "Splitting the Brood".

Super A shallower version of a brood box mainly used for the storage of honey. Can be with or without combs.

Supering The addition of one or more supers to a hive.

Supersedure The raising of a new queen to replace her mother.

Supersedure Cells Queen cells that are built when a colony is preparing to supersede their queen.

Swarm The only reproductive method used by honey bees where the colony produces queen cells, allowing a queen, usually the old one to fly off with a quantity of workers and drones to set up home elsewhere.

Swarm Cells Queen cells that are built when a colony is preparing to swarm.

Uncapping The removal of the cappings of combs of honey prior to honey extraction.

Uniting When two or more colonies are put together.

Workers The incomplete females who make up the majority of the colony. They have many tasks.

INDEX